6/04

Our Affair
with
El Niño

Our Affair with El Niño

How We Transformed an Enchanting Peruvian Current into a Global Climate Hazard

S. George Philander

PRINCETON UNIVERSITY PRESS

PRINCETON AND OXFORD

Library of Congress Cataloging-in-Publication Data

Philander, S. George.
    Our affair with El Niño : how we transformed an enchanting Peruvian current into a
global climate hazard / S. George Philander.
        p.   cm.
    Includes bibliographical references and index.
    ISBN 0-691-11335-1 (cloth : acid-free paper)
    1. El Niño Current.   2. Climatic changes.   I. Title.
GC296.8.E4P485 2004
    551.6 — dc21   2003044235

British Library Cataloging-in-Publication Data is available

This book has been composed in Sabon with Helvetica Narrow & Helvetica Black display

Printed on acid-free paper. ∞

www.pupress.princeton.edu

Printed in the United States of America

10   9   8   7   6   5   4   3   2   1

For Hilda

CONTENTS

ACKNOWLEDGMENTS   ix

PROLOGUE: Assessing Our Affair as It Approaches a
           Critical Juncture   1

PART 1: WHO IS EL NIÑO?

1 A Mercurial Character   11
2 A Fallen Angel?   28
3 A Construct of Ours   34
4 A Matchmaker   40

PART 2: OUR DILEMMA

5 Two Incompatible Cultures   65
6 "Small" Science versus "Big" Science   81

PART 3: COMMON GROUND

7 The Perspective of a Painter   93
8 The Perspective of a Poet   118
9 The Perspective of a Musician   129
10 A Marriage of the "Hard" and "Soft" Sciences   139
11 The Cloud   151

PART 4: A BRIEF HISTORY OF THE SCIENCE

12 Predicting the Weather   161
13 Investigating the Atmospheric Circulation   177
14 Exploring the Oceans   189
15 Reconciling Divergent Perspectives on El Niño   213
16 Taking a Long-Term Geological View   227

PART 5: COPING WITH HAZARDS

17 Famines in India   237
18 Fisheries of Peru   240
19 Droughts in Zimbabwe   244
EPILOGUE: Becoming Custodians of Planet Earth   251

NOTES AND REFERENCES   259
INDEX   273

ACKNOWLEDGMENTS

I am privileged to be a member of the small community of scientists who started studying El Niño when that phenomenon was a curiosity known to few. Countless stimulating discussions and conversations at innumerable meetings of that convivial group of dear friends taught me what I know about El Niño and provided me with most of the ideas presented here. We all agree that the growth of our field over the past few decades has brought enormous changes, but we differ in our assessments of those changes. I assure those who do not share some of the opinions expressed in this book that I look forward to further stimulating discussions. It is a pleasure to acknowledge my esteemed colleagues at GFDL and in the Department of Geosciences at Princeton University, where they have created a most supportive and stimulating environment for studying our planet, its oceans and atmosphere. The detailed comments of Mark Cane, Leo Donner, Bill Lyman, Kelly Sponberg, and Barbara Winter on an earlier draft of the manuscript improved this book enormously and saved me from many errors; those that remain are all my own. John Suppe and Rob Hargraves persuaded me to undertake this project by convincing me that scientists have an obligation to explain their methods, results, and concerns directly to laymen. The talented teaching assistants who alerted me that, to the laymen in Geo 220, scientific facts and theories are less of a challenge than the radically different ways in which we try to solve scientific and social problems include Giulio Boccaletti, Meredith Galanter, Lisa Goddard, Scott Harper, Cara Henning, and Andrew Wittenberg. Michael Bender, Peter deMenocal, Christina Ravelo, and Danny Sigman introduced me to the marvels of paleoclimates. Marta Aizenmann, Laurel Lyman, and Jim Moeller enlightened me about the methods and goals of artists.

Klaus Keller and David Bradford patiently tried to explain to me the mysteries of economics. I completed a first draft of the book during the spring of 2002 while enjoying the generous hospitality of the California Institute of Technology as a Gordon Moore Scholar.

Professor Ed Lorenz, the American Meteorological Society, Cambridge University Press, the University of Washington Press, and *Science* generously gave permission to reproduce certain figures.

Our Affair
with
El Niño

PROLOGUE

Assessing Our
Affair as It
Approaches a
Critical Juncture

"Don't blame me; blame El Niño!"
This slogan, which appears on bumper stickers and T-shirts, attests to the recognition and notoriety El Niño gained in 1997 when he apparently caused innumerable weather-related disasters worldwide. Although everybody now knows of him, he remains a mystery to many people. How did he acquire his name? How is he related to La Niña? If he has been with us for millennia, why did most of us first hear of him in 1997? What was special about that year? How are meteorologists able to anticipate El Niño months in advance when they are unable to predict the weather more than a few days hence? What is the secret of his versatility—his ability to cause floods in California, droughts in Indonesia, mild winters in Canada, the disappearance of fish from the waters off Peru, and the appearance of unusual tropical fish off San Francisco?

These are some of the questions that reporters, journalists, and producers of television documentaries asked me, as a presumed expert, during El Niño's much publicized visit in 1997. This book is my response, written after reflecting on the interviews, which I found both frustrating and rewarding. Frustrating because of the usual difficulties scientists encounter when they try to describe their work to nonscientists; rewarding because I came to realize how marvelously multifaceted El Niño is. Most scientists regard El Niño as simply a puzzle, a natural phenomenon that needs to be described and explained in terms of the laws that govern the motion of the ocean and atmosphere. That narrow perspective makes peripheral the economic and social impacts of El Niño and also the cultural, historical, and political factors that influence scientific research on El Niño. El Niño is so multifaceted that it can be of assistance with a very serious problem we face, bridging the gulf between the worlds of science and of human affairs.

Today we regard El Niño as a peril, a global climate hazard that adversely affects millions of people worldwide. Originally the term El Niño referred to a very different phenomenon, a modest, warm, seasonal current that appears along the shores of Ecuador and Peru around Christmastime, which was welcomed as a blessing. This remarkable change in our perception of El Niño occurred even though that natural phenomenon has remained essentially constant. We changed our views because of changes in our science, politics, economics, sociology, and culture over the past few decades. It is as if we, passionate and capricious, have been having a turbulent affair with steadfast, aloof El Niño! Our relationship is rapidly approaching a critical juncture because constant El Niño is about to become fickle in response to activities of ours. We are provoking a response from him by rapidly changing the composition of the atmosphere. By increasing the atmospheric concentration of carbon dioxide we are inducing global climate changes that soon will include an altered El Niño.

The prospect of an altered El Niño is alarming because we are having more and more trouble coping with familiar El Niño. Our difficulties increase steadily because of a paradox: as we grow in wealth and in population, so does our vulnerability to natural hazards. (Insurance companies are concerned about a steep rise in claims related to damage inflicted by natural hazards, but there is no evidence of an increase in the number and intensity of severe storms, hurricanes, floods, earthquakes. . . .)[1] Scientists are trying to assist us by predicting some of these hazards, thus giving us time to prepare. They are able to do so because of remarkable technological developments and impressive scientific advances over the past few decades. For example, meteorologists have transformed weather prediction from an augury into a reliable source of important information, a splendid achievement that most people take for granted. Now scientists are turning their attention to the prediction of longer-term changes in atmospheric conditions. Their goal is to forecast whether next summer will be warmer and wetter than usual, and whether the following winter will be exceptionally mild or severe.

A first step toward the goal of long-term climate forecasts is the prediction of El Niño. That phenomenon is neither strictly oceanic nor strictly atmospheric but is attributable to interactions between the two media. It has therefore been necessary to merge the traditionally separate disciplines of meteorology and oceanography. A further requirement for predicting El Niño is the continuous monitoring of conditions at and below the ocean surface. Into the 1980s oceanographers were still gathering much of their data from solitary vessels while navigating by means of stars and sextants. (Asking for a "tall ship, and a star to steer her by," was not simply a romantic wish but a practical necessity.) Today satellites in space can pinpoint the position of a ship and can serve as platforms from which we measure conditions at the ocean surface, globally, within hours. Instruments in space are unable to "see" below the ocean surface. To monitor subsurface conditions in the vast tropical Pacific Ocean, oceanographers therefore started to maintain a large array of unattended instruments moored to the ocean floor. Scientists made such rapid progress that, although El Niño caught them by surprise in 1982, by 1997 they could anticipate the arrival of that phenomenon several months in advance. This was an impressive scientific achievement, but it had a most unfortunate blemish.

During the summer of 1997 scientists alerted Californians, on television and in newspapers, of a high probability for exceptionally heavy rains during the winter of 1997–1998 because of a very intense El Niño. Scientists also advised the people of Zimbabwe in Africa that rainfall there would probably be below normal. Californians did indeed experience floods and were prepared, but Zimbabweans had normal rainfall and were unprepared. Because of the expectation that crops would be poor and that farmers would be unable to pay back any loans, banks in Zimbabwe declined loans to farmers. The consequences were dire: crop production was 20 percent below normal in the impoverished country.[2] The prediction of a drought in effect caused a drought.

The tragedy in Zimbabwe raises many questions. Why did the policy makers in Zimbabwe assign too much weight to the scientific predictions? Did they fail to appreciate the significance of a probabilistic forecast? Or did they cynically welcome the forecast of a mysterious threat from the remote Pacific Ocean as an effective means for diverting attention from serious, local political problems? What motivated scientists to make forecasts for Zimbabwe in 1997? Concern for the people of Zimbabwe was of course a major factor, but to what extent were the scientists responding to pressure from their sponsors to demonstrate that their results could be useful? During the Cold War scientists studying El Niño enjoyed remarkable freedom to do science mainly for the sake of understanding an intriguing natural phenomenon, but since the 1980s the sponsors of scientific research have insisted that the focus be on research with practical benefits. For science to flourish, scientists must have a skeptical attitude toward their own results, constantly questioning and testing apparent solutions to problems. They have to adopt a very different attitude when trying to persuade potential clients that their results are useful. How are scientists coping with this predicament? To what degree are such difficulties contributing to the poor communication between scientists and nonscientists?

Ignorance of the *1001 Things Everybody Should Know about Science*[3] is usually cited as the reason for tensions and misunderstandings between scientists and nonscientists. If that were indeed the case, then we could minimize the likelihood of disasters such as the one in Zimbabwe by insisting that policy makers take courses in probability theory. However, the problem has another, more serious dimension. I learned about its scope from teaching my undergraduate course on weather and climate (Geosciences 220 at Princeton University.)[4] The students in this course — economists, historians, politicians, humanists, scientists, engineers, and a few older folks from the town — have diverse backgrounds, but they all share a keen interest in weather and related phenomena. Their enthusiasm to learn about storms, fronts, hurricanes, and tornadoes makes it relatively easy to introduce scientific concepts such as the adiabatic lapse rate and the Coriolis force. Few have trouble with these concepts, but many struggle with something far more fundamental: the profound differences between the worlds of science and of human affairs.

The scientific challenge of predicting El Niño, say, is very different from the challenge of mitigating the impact of an imminent El Niño. The methods and skills required to solve the problems encountered in the cold, uncompromising world of science have very limited applicability in the world of human affairs, where compassion is a virtue, compromise a requisite. Each scientific problem has a well-defined solution that can be found by means of universal methods independent of the investigator's race, gender, or religion. A social problem, by contrast, has a multitude of solutions, each with advantages and disadvantages that are weighed differently in different cultures. So radical are the differences between the worlds of science and of human affairs that their demands are sometimes in conflict. A failure to appreciate these differences is one of the main reasons for misunderstandings between scientists and nonscientists.

The purpose of this book is to help improve communication between scientists and nonscientists by taking advantage of the interest everyone takes in El Niño. The book is a collection of essays that amount to nineteen ways of looking at this multifaceted phenomenon. Part 1, which has four chapters, sets the stage, examines the origin of the name El Niño, relates how we transformed an enchanting current into a global menace, and explains, in chapter 4, why El Niño is a talented matchmaker, capable of integrating disparate communities.

The first step toward bridging a gulf is to become familiar with its characteristics, the topic of part 2 of this book. Chapter 5 explores the radically different ways in which we solve scientific and societal problems, and discusses how the different approaches can give rise to serious dilemmas. Some of these predicaments can inhibit the progress of science, whose curiously undemocratic character is the topic of chapter 6. These matters merit the attention of scientists and also of the sponsors of science, the public. At present the latter group is demanding that scientists engage in "outreach" activities and be involved in the policy aspects of environmental problems, for example. In principle this is laudable, but in practice it is fraught with dangers and can do harm in both the scientific and social spheres. This does not mean that scientists should confine their interests strictly to science. A physicist recently observed that scientists have as much interest in the history and philosophy of science as birds have in ornithology, but the story of our affair with El Niño indicates that such a lack of interest can be a serious liability.

The activities of scientists and those of artists, poets, and musicians have significant differences, but they also have striking parallels. A few are explored in part 3 of this book. (Chapter 7 argues that an artist's depiction of an elaborate panorama on canvas in oil has similarities with a scientist's construction of a realistic computer model of weather and climate.)

Weather forecasting is a source of reliable information today, but the skill of meteorologists was far lower in the 1860s, when routine, daily predictions first became available. Were those forecasts, which had enormous uncertainties, of any value? What factors facilitated the subsequent advances in meteorology? What is required for similar advances in our understanding of, and ability to predict, climate fluctuations such as El Niño? Part 4, a brief history of meteorology and oceanography, addresses these questions. It comes as no surprise that the continual testing of ideas and theories is of paramount importance to scientific progress. Weather prediction has advanced significantly because meteorologists are tested frequently — daily. Those who attempt to predict El Niño, a phenomenon that occurs every few years, have had little experience up to now but, in due course, will become more practiced. When we turn to global warming and related climate changes decades hence, we face a serious problem. The available instrumental records are far too short for stringent tests of models that are used for predictions, and such records will continue to be inadequate for a considerable time to come. To test the models we are obliged to turn to the geological record of remarkably different climates in our planet's distant past. That record, the topic of chapter 16, gives a valuable context for the global warming we are starting to experience.

It will be a while before the forecasts of future climate changes, including global warming, are as reliable as weather predictions are today. We are therefore obliged to make policy decisions on the basis of uncertain and incomplete scientific information. How much scientific information do we need to start implementing effective policies? Part 5 of this book concerns a few specific environmental problems that permit considerable progress on the basis of a bare minimum of scientific information. The outstanding example is the way the government of India copes with poor monsoon rains that cause poor harvests. Such events used to contribute to horrendous famines that led to the deaths of millions of people. That no longer happens, not because of advances in the prediction of the monsoons, but mainly because of critical political changes that facilitated the implementation of effective policies. In the case of global warming, which the epilogue discusses, we must similarly guard against the tendency to defer difficult political decisions on the grounds that the available scientific information is inadequate.

Our affair with El Niño is approaching a critical juncture because we are in the process of altering the composition of the atmosphere and hence the climate of this planet. We acquired the technology to do so only recently, over the past century. We are very pleased with the considerable benefits that accompany this development—an enormous increase in our wealth and population—but we are reluctant to accept that our technological prowess has also brought serious responsibilities. Because we are now geologic agents capable of interfering with the processes that make this a habitable planet, we have become custodians of planet Earth. The decisions we make today will affect not only our offspring for many generations to come but also all the other forms of life on this planet. In the conduct of our affairs we need to be wise and responsible stewards of our planet. How should we proceed?

The main message of this book is that the solutions to serious environmental problems will elude us unless we are all aware of, and respect, the profound differences between the worlds of science and of human affairs. Our biggest challenge is to give appropriate weight to the inevitably uncertain scientific information when making policy decisions. This is too serious a matter to be left to specialists such as scientists and economists. It is the joint responsibility of all of us because the policies we adopt reflect our values. For guidance we can draw on our wealth of experience in dealing with a variety of environmental problems in the past. We gained much of that experience during our affair with El Niño. Much can be learned from him. We need to do so in a hurry, before we succeed in changing him.

Part 1 | Who Is El Niño?

ONE

A Mercurial Character

There are more things in heaven and earth,
Horatio,
Than are dreamt of in your philosophy.
—William Shakespeare, *Hamlet*

Stockbrokers on Wall Street mutter "El Niño" when the market is erratic. Commuters in London do the same when the traffic is exceptionally bad. Humorists everywhere depict El Niño as a little devil responsible for everything that goes wrong. The rascal is so remarkably versatile and ubiquitous—the strange weather he causes globally includes floods and droughts, mild and severe winters—that his name has become part of our vocabulary; it designates a mischievous gremlin. El Niño joins a host of meteorological phenomena that serve as metaphors in our daily speech: the president is under a cloud; the test was a breeze; the economy is in the doldrums. The meanings of these statements are perfectly clear because everyone has intimate familiarity with clouds and breezes, and is aware of the depressing, perennially overcast doldrums near the equator. El Niño, however, remains a mystery to most people, despite all the publicity he currently receives. Who is El Niño?

El Niño is Spanish for "the boy" and can also be translated as "the small one." This is an odd name for a global phenomenon that adversely affects millions of people worldwide. Is irony intended? The name becomes even more intriguing when we investigate why the first letters in El Niño are capitalized. Apparently the people of Ecuador and Peru, who first used the name, had in mind, not any boy, but specifically the Child Jesus. Why that name for a phenomenon that amounts to a disaster? Were some of the early South American converts to Christianity cynical? Not at all. Originally the term El Niño referred to a warm coastal current that appears along the shores of Ecuador and Peru around Christmastime, when it brings welcome relief from the cold waters that otherwise bathe those shores.[1] The transformation of a regional curiosity, which we used to welcome as a blessing, into a global climate hazard happened recently, during the second half of the twentieth century.

We first "encountered" him, more than a century ago, along the shores of Ecuador and Peru. We assumed that he was an angel and named him El Niño. We eventually identified his relatives—La Niña, Southern Oscillation, ENSO—and proceeded to devote learned tomes to descriptions and exegesis of this remarkable family. These scriptures provide such a rich spectrum of historical, cultural, and scientific perspectives on El Niño that we are now having difficulties interpreting our own texts. We have become confused about issues as fundamental as the identity of El Niño. A member of our clergy, a scientist, summarizes the current, bewildering state of affairs as follows:[2]

> The atmospheric component tied to El Niño is termed the "Southern Oscillation." Scientists often call the phenomenon where the atmosphere and ocean collaborate ENSO, short for El Niño–Southern Oscillation. El Niño then corresponds to the warm phase of ENSO. The opposite "La Niña" ("the girl" in Spanish) phase consists of a basinwide cooling of the tropical Pacific and thus the cold phase of ENSO. However, for the public, the term for the whole phenomenon is "El Niño."

The clerics are confused. They have been grappling with the question Who is El Niño? for some time, and they find the answer to be frustratingly elusive. In the early 1980s they appointed an august committee to define El Niño quantitatively,[3] but shortly after that committee promulgated a definition, based on the "typical" behavior of El Niño up to that time, El Niño visited and behaved in a manner entirely inconsistent with the definition. His visits used to start along the shores of Peru, whereafter he proceeded westward across the Pacific, but in 1982 he reversed his itinerary: he first appeared in the far western equatorial Pacific and then moved eastward, arriving in Peru in a season different from the "usual" one. His personality proved too complex to be captured with a simple definition involving a handful of numbers. Clergymen who insist on narrow, rigid definitions of El Niño are at risk of becoming Mr. Gradgrind:

> "Bitzer," said Thomas Gradgrind, "your definition of a horse."
>
> "Quadruped. Gramnivorous. Forty teeth, namely twenty-four grinders, four eye-teeth, and twelve incisive. Sheds coat in the spring; in marshy countries sheds hoofs too. Hoofs hard, but requiring to be shod with iron. Age known by marks in mouth." Thus (and much more) Bitzer.
>
> "Now girl number twenty," said Mr. Gradgrind, "you know what a horse is."
>
> Charles Dickens, *Hard Times*

To methodical scientists, who cope best with an idealized, consistent world, El Niño is frustratingly whimsical. He was brief and intense in 1997, mild and languorous in 1992. He can be regular and energetic during some decades (the 1960s) but practically absent during other decades (the 1920s and 1930s). Laymen, who are more tolerant of caprice and impulse than scientists are, find him fascinating and beguiling, as is evident from the humorous ways in which they now use the term El Niño as a metaphor in their daily speech. The clerics should applaud this development because it offers a solution to their quandary.

The use of words cannot be legislated. Today everyone associates El Niño with the appearance of warm waters in the eastern tropical Pacific and believes him to be capable of interfering with weather patterns worldwide. This means that the term El Niño, as used at present, refers to a phenomenon with both atmospheric and oceanic aspects. Those who insist that El Niño is a strictly oceanic phenomenon, who recall that it originally referred to a seasonal current along the coast of Peru, can take pride in their erudition, but they should realize that their use of the term is becoming archaic. Unless they wish to be an elite culture with its own unintelligible argot, they should follow those who accept that El Niño has both atmospheric and oceanic aspects.

Until 1957, a lack of measurements led us to believe that El Niño was a regional phenomenon, confined to the shores of Peru and Ecuador. When he visited that year, many scientists happened to be engaged in a program to collect atmospheric and oceanic data on a global scale over an extended period.[4] The data revealed that we had been completely mistaken about the true dimensions of El Niño. We discovered that he is associated with the appearance of unusually warm waters, not only along the western coast of South America, but right across the vast tropical Pacific. It was natural to conclude that such an exceptional state of affairs amounts to a major departure from "normal" conditions. We therefore analyzed El Niño's subsequent appearances in order to identify the "triggers" that initiate the development of such unusual conditions. At first some scientists proposed that a collapse of the trade winds precedes the appearance of El Niño.[5] Today many believe that the "triggers" are bursts of westerly winds along the equator in the vicinity of the date line; critical subsequent developments include oceanic Kelvin waves that propagate eastward along the equator. (The detection of these waves in satellite photographs of the Pacific is sometimes regarded as an annunciation, to be celebrated with a press conference.)

A very different perspective on El Niño became available in the 1980s once we had records sufficiently long to cover many consecutive episodes. To share this perspective, inspect surface temperature fluctuations in the eastern tropical Pacific over the past century. (The occasional warming of the surface waters in that region amounts to El Niño's signature.) Figure 1.1 calls into question the earlier view that El Niño is a departure from "normal" conditions. Such conditions, which correspond to the horizontal line, are seen to prevail very seldom. Temperatures are either above the horizontal line, during El Niño episodes, or they are below that line. We are dealing with an unending oscillation that has a distinctive timescale of approximately four years. It is as if we are listening to a continuous melody that has a distinctive

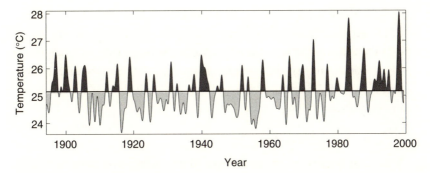

Figure 1.1. Temperature variations, in degrees centigrade, over the past 100 years, as measured on the equator near the Galápagos Islands after filtering out of the seasonal cycle. Warm conditions correspond to El Niño, cold conditions to La Niña.

beat. El Niño is only part of this ever present tune, so it makes little sense to listen to his notes in isolation. To appreciate his high, shrill notes we should also pay attention to the deep, resonant notes of the complementary periods when the waters are cold. La Niña is an apposite name for those cold periods.[6] The continual Southern Oscillation between El Niño and La Niña seems to have no beginning or end.

Are El Niño and La Niña the complementary phases of a cycle with no beginning or end? Or is each one of them an independent event, a departure from "normal" conditions "triggered" by certain disturbances, some of which lead to the appearance of warm water, some to the appearance of cold water. These different points of view reflect different perceptions of time itself. Those who believe that "triggers" initiate El Niño regard time as an arrow that moves in a definite direction as El Niño progresses from a beginning to an end, from birth to death. A more reassuring perception of time, one that gives us a continual sense of renewal, is in terms of natural cycles. El Niño can then be regarded as part of an endless cycle, similar to those mentioned in Ecclesiastes:

The sun also ariseth, and the sun goeth down, and hasteth to his place where he arose. The wind goeth toward the south, and returneth about unto the north; it whirlith about continually, and the wind returneth again according to his circuits. All the rivers run into the sea; yet the sea is not full; unto the place from whence the rivers come, thither they return again. . . . The things that hath been, it is that which shall be; and that which is done is that which shall be done.

Ecclesiastes 1:5–9

Much of literature is concerned with these two very different aspects of time, which are captured by the metaphors of time's arrow and time's cycle. The current debate about which of these metaphors best describes El Niño is reminiscent of the nineteenth-century debate about the interpretation of the geological record.[7] Some early geologists, who subsequently became known as "catastrophists," described the history of the earth chronologically, in terms of a sequence of mostly biblical events and catastrophes that moved from a definite beginning to the present. (James Ussher, while he was bishop of Armagh in Ireland, determined that the creation started at precisely 9:00 A.M. on Monday, October 23, 4004 B.C.) In 1795 James Hutton put forward a radically different perspective when he proposed that the geological record extends back over an inconceivable length of time and should be interpreted in terms of repeated cycles with "no vestige of a beginning, — no prospect of an end."[8] Hutton believed that unchanging geological processes such as erosion and the gradual uplifting of rocks, acting slowly and steadily over an immensity of time that is difficult to comprehend, shaped our landscape in the past and continue to do so today. His followers were therefore known as "uniformitarians." Although they developed convincing arguments that explain much of the geological record, it is difficult to deny that the fossil record tells a story of the sequential evolution of different species, a story that moves in a direction. After much debate, geologists reached an accommodation that has room for both time's arrow and time's cycle. Students of El Niño need to do the same.

Persuasive evidence that, in 1997, a burst of westerly winds contributed to the development of El Niño lends credence to the idea that El Niño has a definite beginning. However, similar wind bursts on other occasions have failed to produce El Niño. Apparently only bursts that appear at the "right" time are capable of inducing El Niño. What factors determine the "right" time? Consider a swinging pendulum subjected to modest blows at random times. A blow at the right time can increase the amplitude of the swing considerably. At the wrong time, it can cause the pendulum to come to a standstill. This argument suggests that we are dealing with an unending cycle, subject to random disturbances. This compromise between time's arrow and time's cycle explains why each El Niño is distinct. The phenomenon was particularly intense in 1997 because, as it was about to visit, a burst of westerly winds came along, causing a significant amplification and acceleration of developments. The absence of appropriate random disturbances is the reason why El Niño was weak and prolonged in 1992.

We have seen the signatures of El Niño and La Niña in figure 1.1, which tells us when each of them visited, but we still do not know what those phenomena look like. What are their distinctive features? Their pictures, in figure 1.2, are surprising and also sobering because they bring to mind Confucius' observation that "a common man marvels at uncommon things; a wise man marvels at the commonplace." We have been lavishing attention on uncommon El Niño, whom some of us regard as a departure from "normal" conditions, when in reality commonplace La Niña is the more interesting of the two! El Niño's temperature patterns at the ocean surface are downright plain; uniformly warm surface waters are exactly what we expect in the tropics where sunshine is most intense. La Niña, by contrast, is intriguing and mysterious: she remains admirably cool under intense sunlight and expresses her coolness with flair. Her sea surface temperature patterns have fascinating asymmetries. Although the intensity of sunlight is independent of longitude and is perfectly symmetrical about the equator, she keeps the waters of the tropical Pacific colder in the east than the west and, in the east, warmest in a band to the north of the equator. La Niña transforms that renowned line, the equator, from a mere geographer's artifice into

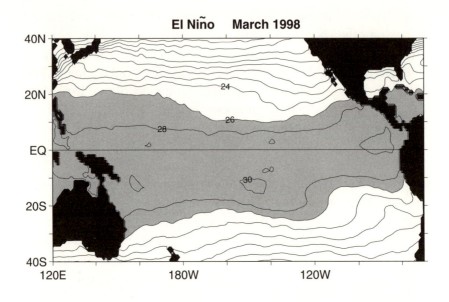

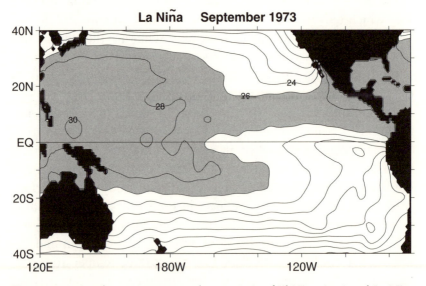

Figure 1.2. Sea surface temperatures characteristic of El Niño (*top*) and La Niña (*bottom*). Aspects of the seasonal cycle are evident in these snapshots, which show temperatures at the end of the Southern Hemisphere summer (*top*) and at the end of the Northern Hemisphere summer (*bottom*).

something very special — the location of a westward-stretching sliver of exceptionally cold waters, rich in nutrients and hence in marine life. So plentiful is nourishment along the Line that Moby Dick used to loiter there. That is where Captain Ahab went looking for the white whale.

La Niña expresses her alluring asymmetries not only in oceanic conditions but in atmospheric conditions too. The distinctive pattern that describes the sea surface temperatures associated with La Niña also describes the rainfall and cloud distribution associated with that phenomenon. For example, the dreary doldrums coincide with the swath of warm water that stretches clear across the Pacific along the latitude of approximately 10° N. That zone and, more generally, regions of high sea surface temperature in low latitudes have plentiful rainfall because, in low latitudes, the moist air rises spontaneously into tall cumulus towers over the warmest waters. The air subsides over regions of cold surface waters, where the clouds are very different; they are low, horizontal stratus decks that provide no precipitation and that stretch far westward off the coasts of California and Peru. During El Niño, when warm waters cover the eastern tropical Pacific, the stratus clouds disappear from that region and are replaced by cumulus clouds that bring heavy rains. The coastal zones of Ecuador and Peru can now experience floods while rainfall is reduced in the far western equatorial Pacific; northern Australia, New Guinea, and the Philippines can have droughts at such times.

How does La Niña succeed in keeping the equator cold despite the intense sunlight at that latitude? By bringing the winds into play, using them to bring the cold waters of the deep ocean to the surface. This is accomplished by tilting the thermocline, the interface between the very warm waters of the upper ocean and the much colder water at depth. To a first approximation, the tropical oceans are composed of two layers of fluid: the upper one is warm and shallow, the lower one cold and very deep. The intense westward winds that characterize La Niña pile up the warm water in the west, causing the thermocline to tilt steeply so that cold water is exposed to the surface in the east. During El Niño, when the winds are relaxed, the interface between the warm surface

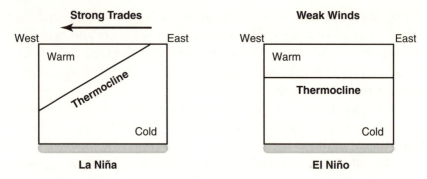

Figure 1.3. A schematic diagram of changes in the slope of the thermocline from La Niña (*left*) to El Niño (*right*). Note that, in the east, cold water is exposed to the surface during La Niña.

waters and the colder water at depth becomes practically horizontal. (See figure 1.3.)

Changes in the winds change the slope of the thermocline and thus alter surface temperature patterns in the tropics. This is a tantalizing result because the converse is also true: changes in temperature patterns change the winds. (A familiar example is a sea breeze from the cool ocean toward the warm land during the day; the wind blows in the opposite direction at night, when the land is cooler than the ocean.) The absence of temperature differences between the western and eastern tropical Pacific during El Niño explains why he is associated with relaxed winds. La Niña, on the other hand, has large temperature differences and hence strong winds. According to this circular reasoning, wind fluctuations are both the cause and consequence of changes in sea surface temperature patterns! From this chicken-and-egg argument scientists infer that interactions between the ocean and atmosphere are at the heart of the matter. Those interactions can amount to an escalating tit for tat: a slight intensification of the winds during La Niña increases the temperature difference between the eastern and western equatorial Pacific, thus reinforcing the winds, so that the temperature difference increases even further. . . . (Scientists and engineers refer to such interactions by the unattractive term "positive feedback.") These intimate interactions between the ocean and atmosphere, on the grand scale of the vast Pacific Ocean, generate El Niño and La Niña. They are therefore the children of water and air.

Laymen take the lineage of El Niño and La Niña for granted and accept that the two have both atmospheric and oceanic aspects. It then follows that, because the two together constitute the Southern Oscillation,

El Niño + La Niña = The Southern Oscillation

all three phenomena must involve both the air and the sea. This impeccable logic makes the acronym ENSO (= El Niño + Southern Oscillation) tautological and unnecessary. Use of the term would be justified if El Niño were merely an oceanic current, if the Southern Oscillation were a strictly atmospheric phenomenon. We held such beliefs at one time, but no longer. The continued use of the term ENSO is therefore a puzzle. It seems to signify different things to different authors and thus causes confusion.[9] Could its use reflect the determination of the clergy to have their own jargon?

Neither the ocean by itself nor the atmosphere on its own is capable of producing a Southern Oscillation—the one needs the winds, the other the surface temperature changes. However, the ocean plus atmosphere in concert, the two coupled together to form an inseparable unit, can spontaneously generate a Southern Oscillation. The properties of that oscillation depend on a key difference between the ocean and atmosphere: whereas the ocean is very slow in responding to changes in the winds, the atmosphere is swift in adjusting to altered ocean temperatures. We can simulate the interactions between the ocean and atmosphere by taking a shower in a bathroom with old-fashioned plumbing. When we turn the shower on, the temperature of the water at first is too cold. We therefore turn the knob of the faucet toward warm. Because the sluggish plumbing takes a while to respond, the temperature increases so slowly that we impatiently continue turning the knob beyond warm toward hot. After a while the water is pleasantly warm, but it then becomes progressively hotter and hotter so that we start turning the knob back toward warm and even toward cold. In due course the water is too cold, and we once again turn the knob toward warm. A seesaw between temperatures that are alternately too warm or too cold continues indefinitely because we respond instantaneously to the temperature of the water, while the plumbing responds to our instructions in a delayed mode.

In this analogy, the person in the shower represents the quick and nimble atmosphere; the plumbing corresponds to the ponderous ocean. In the same way that interactions between the person and the plumbing are essential to the oscillation in the water temperature, so interactions between the atmosphere and ocean are essential to the Southern Oscillation between El Niño and La Niña. The delayed response of one partner to a move by the other — of the ocean to a change in the winds — determines the period of the oscillation. These insights into the factors that determine the timescale of the Southern Oscillation, and an understanding of how El Niño and La Niña are related, are the rewards for focusing on time's cycle. In an idealized world with no beginning or end, the cycles are perfectly symmetrical; one El Niño is indistinguishable from another, and each is the mirror image of La Niña. Furthermore, the music of this pair is a monotone with a perfectly steady beat, so that we might as well be dealing with a pendulum that monotonously swings to and fro in a perfectly predictable manner. To approach complex reality, where each El Niño is distinct and where the music is a lively tune with a syncopated rhythm, we have to accommodate time's arrow and introduce brief, sporadic westerly wind bursts. They contribute to the relatively rapid, high-frequency aspects of the music, the notes played by flutes and violins. In addition, there are gradual changes in the properties of El Niño from one decade to the next. To explore that low-frequency aspect of the music, which involves a cello or double bass, we once again examine the signatures of El Niño and La Niña. In figure 1.1 we distinguished between El Niño and La Niña by adopting as a reference line the horizontal axis, which represented "normal" conditions, the average temperature for the past one hundred years. That line in effect defined the key (or pitch) of the Southern Oscillation's music. A tune that is always played in the same key soon becomes boring. Innovative musicians add color and variety to their music by occasionally modulating to another key, playing the tune at a higher or lower pitch. Are we underestimating the subtlety of El Niño's music by arbitrarily assuming that it is played in an unchanging key? A reference line on the basis of data collected over the past one hundred years is an arbitrary choice. Why not the past two hundred years, or the past decade? In figure 1.4 the upper panel has a constant reference line—that panel is identical to figure 1.1—but the lower panel has a reference line that changes continually because it corresponds to the average over a relatively short period of 10 years. The superimposed, rapidly fluctuating temperature curves are identical in the two panels, but El Niño's signature looks different in each one because the key in which the music is played is different. It is constant in the upper panel but shifts continually and gradually in the lower panel.

In the upper panel, El Niño attained exceptionally large amplitudes on two occasions, in 1982 and in 1997, and was exceptionally persistent during the early 1990s; La Niña seemed to disappear during the 1980s and 1990s. The lower panel tells a different story: El Niño is seen to alternate with La Niña throughout the record, even during the 1990s; the prolonged El Niño that started in 1992 now simply amounts to the persistence of background conditions. This revised interpretation results from the adoption of a new reference line that makes explicit that El Niño is superimposed on a continually changing background state, a gradual warming and cooling of the eastern Pacific over many decades.[10] It is as if the pitch of the music were gradually and continually changing. Such modulations to other keys provide a musician with new opportunities, new chords for harmony, for example. El Niño too explores new possibilities when the pitch of his music changes. The gradual, decadal warming and cooling of the tropical Pacific has subtle influences on El Niño. Careful inspection of his signature indicates that he visited every

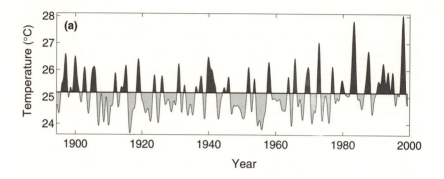

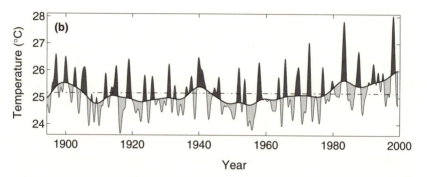

Figure 1.4. The top panel is the same as fig. 1.1. The bottom panel shows the same temperature fluctuations, but now the reference line, instead of being the average temperature of the past 100 years, changes gradually and is, at each point, the average temperature over a 10-year period.

three years, approximately, during the relatively cool 1960s and 1970s but then, during the warmer decades of the 1980s and 1990s, increased the interval between his visits to about five years. A change to a lower pitch was accompanied by a change in the rhythm, from fast to slow. Furthermore, after the warming in the late 1970s, El Niño's journeys across the Pacific proceeded from west to east, whereas, during the 1960s and 1970s, there were occasions when the journeys were in the opposite direction. Despite these changes, El Niño and La Niña remain recognizable. The complex music has a main theme that can readily be identified, thus providing unity within variety: the eastern tropical Pacific warms during El Niño, cools during La Niña.

Versatile El Niño can produce music with infinite variety because he has many tricks up his sleeve. To prevent an oscillation of the coupled ocean-atmosphere, with a period of several years, from becoming a dull monotone, El Niño can rely on gradually changing background conditions that, from one decade to the next, change the pitch of the music; or he can call on brief, sporadic westerly wind bursts that amplify or diminish, shorten or prolong the warm conditions. Scientists have yet to identify the additional devices by which La Niña lengthens her stay while El Niño curtails his. (In 1982 and again in 1997 El Niño conditions lasted for about one year, whereafter La Niña conditions persisted for about four years.) Scientists do not know why the oscillation between these two states can become skewed in this manner, which suggests that El Niño is likely to continue surprising them in future.

El Niño emerges from interactions between the tropical Pacific Ocean and the atmosphere above it. So intense are those interactions that regions remote from the tropical Pacific are affected. For example, El Niño can contribute to heavy rains over California and the Gulf states and to mild winters in central Canada and the northeastern United States. If El Niño reaches his peak during the northern summer, then the likelihood of more hurricanes in the Atlantic increases. To the west of the Pacific it is possible for El Niño to contribute to reduced rainfall over India and over Zimbabwe in southeastern Africa while increasing rainfall over equatorial eastern Africa.[11]

El Niño is capable of affecting different parts of the globe, but we should guard against exaggerating these "teleconnections" that emanate from the tropical Pacific. For example, California can have heavy rains even in the absence of El Niño. Mild winters in the northeastern United States and central Canada are more likely when El Niño visits, but such winters are possible even when he is absent. Although El Niño can contribute to reduced rainfall over India and Zimbabwe, those countries had normal rainfall during the very intense event of 1997–1998. The point is that atmospheric conditions in different parts of the globe depend on a great many factors; El Niño is only one of them. Relatively few instances of "unusual weather" can be attributed to him. (His influence over northern Europe is negligible.)

A close relative of El Niño makes an occasional appearance in the tropical Atlantic. This should come as no surprise, because the climates of the tropical Atlantic and Pacific have striking similarities. Westward trade winds prevail over both oceans, creating similar sea surface temperature patterns with warm waters along the western rims, cold waters in the east, except in a narrow band of latitudes just north of the equator. The western shores of Africa resemble those of the Americas in having barren deserts adjacent to the cold oceans, except for lush vegetation in the band just north of the equator, where the seas are warm and rainfall is plentiful. A rise in sea surface temperatures in the eastern side of either ocean causes a relaxation of the trade winds over that ocean and brings rains to the otherwise arid shores of southwestern Africa in the case of the Atlantic, Peru in the case of the Pacific. Such occurrences are more sporadic, and less intense, in the Atlantic than in the Pacific. In the latter ocean, El Niño is one phase of the irregular, continual Southern Oscillation, which can be compared to a freely swinging pendulum. In the Atlantic the pendulum is strongly damped and hangs vertically for much of the time, infrequently making a single swing; those are the occasions when El Niño makes an appearance, often in the wake of La Niña. To scientists, this difference between the Atlantic and Pacific amounts to a valuable test for their theories. Can they explain why the Southern Oscillation is strongly damped in the Atlantic but not the Pacific? Presumably the answer involves the smaller dimensions of the Atlantic. El Niño is inhibited on the cramped stage of the Atlantic but has ample room to develop to his full potential in the vast Pacific. Because of its modest amplitude, the Atlantic phenomenon is a regional one that affects mainly the adjacent land areas, whereas its far more extensive Pacific counterpart has a more global impact.

For most of the time, the Atlantic and Pacific Oceans experience La Niña conditions—the cold regions off South America warm up only when El Niño visits. The Indian Ocean is intriguingly different; there El Niño conditions are prevalent. The eastern part of that ocean is warm most of the time and cools off infrequently, during sporadic La Niña episodes. This happened in 1997 when westward winds prevailed along the equator and brought exceptionally heavy rains to Kenya. It is no coincidence that El Niño was visiting the Pacific at that time; his presence there sometimes favors complementary conditions in the Indian Ocean.[12]

El Niño is such a chameleon, adopting a different shape each time he visits the Pacific and whenever he appears in the other tropical oceans, that attempts to define him narrowly should be abandoned. We need to accept that the term El Niño is useful in the same way that the term winter is useful even though each winter is distinct. No one wishes for strict, quantitative definitions of winter and summer after reading the following lines:

> Now is the winter of our discontent,
> made glorious summer by this sun of York.
> William Shakespeare, *Richard III*

Hopefully a poet will soon use the term El Niño so imaginatively that the cognoscenti will stop debating its precise definition.

TWO

A Fallen Angel?

Some are born great, some achieve greatness,
And some have greatness thrust upon 'em.
—William Shakespeare, *Twelfth Night*

Every year, at Christmastime, a modest, warm southward current appears along the coast of Ecuador and Peru, providing such a welcome respite from the frigid waters that usually bathe those barren shores that we named it El Niño, after the Child Jesus. Sometimes the adorable infant lingers longer than usual, brings heavy rains, and bears exotic gifts. A visitor to Peru described one such joyous occasion, in 1891, as follows:

> the sea is full of wonders, the land even more so. First of all the desert becomes a garden. . . . The soil is soaked by the heavy downpour, and within a few weeks the whole country is covered by abundant pasture. The natural increase of flocks is practically doubled and cotton can be grown in places where in other years vegetation seems impossible.[1]

At such times El Niño fills the sea with "wonders" that, on occasion, have included long yellow-and-black water snakes, bananas, and coconuts. These *años de abundancia* (years of abundance) used to steal our hearts, to such a degree that we started reserving the name El Niño for them. (Today the term refers to these interannual episodes and no longer to the annual current.)

Over the years, the child grew into a mischievous youngster whose pranks included frightening off the fish that are plentiful when the waters off Peru are cold. The huge bird population that lives off these fish produces guano so abundantly that, in the middle of the nineteenth century, this exceptionally fine fertilizer was mined on certain islands off the coast of Peru, for export to farmers in Europe and northern America. The large amounts of money Peruvians earned in this manner dwindled to nothing when, after a few decades, the resource was exhausted. El Niño was implicated, albeit in a minor way; he contributed to the disappearance of the fish and hence to a decrease in the number of birds, thus causing the guano to be replenished more slowly.

El Niño's transition to adulthood was abrupt. It happened in 1957. When he visited that year we discovered that the youngster was no longer a modest coastal current off Peru but had grown into a phenomenon of the entire tropical Pacific Ocean, and the global atmosphere too! Overnight he acquired the ability to interfere, not only with the fisheries of Peru, but also with global weather patterns, including the monsoons over India. His potential for mischief was now so enormous that we resolved to watch him closely.

In the 1960s and 1970s several nations formed consortia to research El Niño and his relatives in the tropical Pacific Ocean and in the Atlantic and Indian Oceans too. We enhanced our understanding of the coastal currents off Ecuador and Peru by studying those off California and Somalia. We paid special attention to the equator once we discovered that that renowned Line is a favorite haunt of El Niño. There he consorts with swift equatorial jets, mysterious equatorial undercurrents, and curious equatorial waves that use the Line as a conduit to speed across the oceans.[2] We accumulated detailed descriptions of El Niño's various visits and distilled from them the morphology of a canonical visit. To make sense of our diverse observations, we developed theories and models that exploited the speedy computers that were becoming available. Our progress was rapid, and by the early 1980s we thought that we had his measure. Then, in 1982, he confounded us.[3]

El Niño, on his previous visits, had first appeared along the shores of Peru in March and then proceeded westward. In 1982 he caught us by surprise by changing his itinerary. That year, he was first sighted in the remote western Pacific Ocean, whereafter he proceeded eastward. As a result, he arrived in Peru, not during Christmas as usual, but far earlier. Such capricious behavior can be dismissed as the rebelliousness of an adolescent, except that it caused devastation on an unprecedented scale. We therefore renewed our vigil, but only after recognizing that El Niño has a companion.

Up to this time, we tended to think of El Niño as an isolated phenomenon, a departure from "normal" conditions, but it now became clear that he is part of a continual oscillation. An oscillation is usually between two complementary states—winter and summer in the case of the seasonal cycle. What is an appropriate name for El Niño's consort, his complement? An early proposal, not in use for long, was the unfortunate if not apocalyptic term anti–El Niño. Another suggestion, El Viejo, or the Old One, presumably appealed to those who believe in reincarnation; it did not prove popular. The somewhat frivolous name that everyone finally adopted is La Niña. She usually precedes El Niño and returns again after his departure. What processes sustain this continual Southern Oscillation between El Niño and La Niña? Why is it irregular? How far in advance can El Niño be predicted?

To address these questions, we embarked on a massive ten-year international research program to describe, explain, and predict El Niño. We installed a huge array of instruments to monitor the vast Pacific Ocean and launched several satellites, some geostationary, some polar orbiting, to watch from space. This information, plus improvements in the models, led to a better understanding of, and an apparent ability to predict, El Niño. However, it soon was evident that although we had become better acquainted with El Niño, he had become more wily. No one anticipated that, in 1992, El Niño would linger for an exceptionally long time—several years—before gradually fading away. In 1997 he brought more surprises. The forecasters anticipated that he would visit that year but failed to forecast how intense he would be. He wreaked havoc over large parts of the globe but also made several generous gestures: northern America enjoyed a mild winter; in India and southeastern Africa El Niño refrained from decreasing the rains the way he often does.

As El Niño grew in size, power, and guile, so did our infatuation with him. We adored the child, doted on the youth, and became obsessed with the man. Today, countless documentaries on television and radio, and numerous articles in newspapers and magazines that range from *Reader's Digest* to *Soybean Digest,* keep everyone informed of each and every move he makes. Analyses of his impact on the global economy abound: by reducing the harvests of coconuts in the Philippines and of anchovies off Peru, he increases the prices of soaps and detergents with coconut oil as an ingredient, of fish meal fed to chickens, and of soybeans that can serve as a substitute for fish meal. Directly or indirectly, El Niño affects all of us. That is presumably the reason why we lavish more and more attention on him. Unfortunately, he responds by behaving worse and worse. The damage he inflicted during his 1997 visit has been assessed in the following stark terms:

Number of deaths attributed to El Niño of 1997	21,706
Number injured and physically affected	117,862,114
Number of people displaced and made homeless	4,829,884
Damage inflicted, globally, in U.S. dollars	$33 billion[4]

And yet we continue calling him El Niño, a term of affection. There must be more to the story. Did the government officials and insurance agents who estimated the damage inflicted by El Niño during his 1997 visit take into account the favors he did us that year? (The northeastern United States enjoyed a mild winter that reduced heating bills, provided a boon to the construction industry, and decreased the number of people suffering heart attacks while shoveling snow.)

Many of the benefits we derive from El Niño are intangible. For example, he assists with bridging the gulf between the "two cultures" of science and the liberal arts by rekindling the interest of laymen in science. Today many people are disillusioned with science because it seems to concern mainly phenomena that we cannot perceive with our senses—unimaginably small particles, inconceivably huge clusters of galaxies. El Niño helps us remedy this problem because he makes clouds, winds, rain—things that are so endlessly fascinating and dear to us that they are the subjects of our poems—even more captivating. He teaches us that changes in weather and climate are not merely local phenomena that profoundly influence the traditions and culture of the people in a region; they are aspects of continually changing global patterns that connect all of us. (We glimpse those patterns in the swirling white clouds that envelop our fragile blue planet in pictures taken from space.) El Niño reveals a unity behind apparently disparate phenomena and demonstrates vividly that a change in climatic conditions anywhere has an effect everywhere; his signature is a warming of the remote eastern tropical Pacific, but his impact is global. El Niño thus provides scientists with opportunities to counter those who contend that natural disasters, rather than being phenomena with rational explanations, amount to punishment for our sinful ways. Furthermore, by translating an understanding of El Niño into an ability to predict that phenomenon, scientists can demonstrate to paternal politicians the importance of "basic" research. (Those politicians insist that scientific research be of direct and obvious benefit to society, despite numerous examples of seemingly abstruse scientific results that subsequently were translated into technological marvels.)

From El Niño we can learn, not only about science, but also about the practice of science. To cope with the great diversity of natural phenomena, science has separate disciplines such as meteorology, oceanography, geology, biology. Nature is unaware of these categories so that some phenomena fall between the cracks. El Niño is a splendid example—it can be explained in neither strictly atmospheric nor strictly oceanic terms because it is the product of interactions between the ocean and atmosphere. El Niño should therefore be credited with arranging a successful merger of meteorology and oceanography, two traditionally separate and independent fields. (We need only consider our very different attitudes toward the atmosphere and oceans to appreciate why the success of this marriage is a major achievement.)

In summary, El Niño's behavior is reprehensible at times, but he remains lovable because his deeds implicate nobody. This is in sharp contrast to global warming, which has the annoying habit of indicting marvelous inventions that contribute to our well-being and high standard of living, the combustion engine for instance. It should come as no surprise that global warming is controversial and divisive. El Niño, on the other hand, is adorable and endearing. He is involved with phenomena, weather and climate, with which we have such intimate familiarity that they are topics of our daily conversation. Are we sure that he is a fallen angel?

THREE

A Construct of Ours

When I was growing up in Peru I never heard of El Niño. Amazing!

—Jorge Sarmiento (remark made during a seminar in Princeton, N.J., on November 10, 2002)

Although El Niño has been with us for many millennia, most people first heard of that phenomenon in 1997, when it joined the ozone hole and global warming as major environmental problems of concern to everyone. Whereas the ozone hole over Antarctica appeared unexpectedly in the 1980s, and global warming is a potential problem, El Niño has always been with us, in essentially the same form. Why then did we not recognize it as a hazard until very recently?

The developments that set the stage for the enormous publicity El Niño received in 1997 occurred throughout the tumultuous twentieth century and include not only astounding scientific and technological advances but also enormous growth in our wealth and in our population. Science revealed that what we at first regarded as a regional climate fluctuation confined to the shores of Ecuador and northern Peru is but part of a phenomenon that affects the entire globe. That apparent growth in the physical dimensions of El Niño seemed to accompany a growth in his destructiveness. This perception is a consequence of the increase in our prosperity, especially after World War II, a period of unprecedented economic growth in the world economy. Although this "Golden Age" benefited the developed capitalist countries the most, practically all nations raced ahead.[1] The reasons why people originally adored El Niño are still there—he still brings rain to certain arid parts of Peru, converting the desert into a garden—but we now pay scant attention because we are too concerned about the roads, bridges, and homes that the rains wash away. Today, when El Niño visits, the sea off Peru is still "full of wonders," but we no longer consider those wonders adequate compensation for the temporary disappearance of the usually abundant cold-water fish. Catching those fish has become a major industry that is dealt a crippling blow when El Niño visits. In Peru and elsewhere, the severe storms, tornadoes, and hurricanes that we now attribute to El Niño inflict far more damage today than similar phenomena did in the past, not because those hazards are more frequent or more intense, but because we have more possessions that can be damaged, and because too many of us choose our places of abode unwisely.[2] Rather than acknowledge that this is the case, we find it more convenient to make El Niño a scapegoat for problems mostly of our own making. As a consequence we have transformed him from a blessing into a menace.

During the decades after World War II fisheries in Peru grew so rapidly that, by the 1970s, the fish stock was endangered. When El Niño visited in 1972, the fisheries crashed, causing local economic devastation. By that time many scientists, but very few laymen, had come to the realization that El Niño is a global rather than regional phenomenon. The public at large remained unaware of this new perspective until 1982, when El Niño was so exceptionally intense and inflicted such extensive damage, worldwide, that the phenomenon received considerable publicity in newspapers and on television. Numerous articles described how seemingly unrelated phenomena in different parts of the globe—droughts in the Philippines and floods in California—all had a common cause, high sea-surface temperatures in the eastern tropical Pacific.

Climate is usually regarded as a regional phenomenon that strongly influences local culture. El Niño of 1982 drew attention to the intriguing notion of global climate changes and set the stage for renewed interest in a separate phenomenon, one the Swedish chemist Arrhenius had first discussed almost a century earlier. Arrhenius warned that our industrial activities would lead to an increase in the atmospheric concentration of carbon dioxide and that the result would be global warming. Initially scientists disregarded this prediction on the grounds that the oceans would absorb the carbon dioxide we release into the atmosphere. Only in the 1960s did they establish that the atmospheric concentration of carbon dioxide is indeed rising rapidly because of our activities and has been doing so since the start of the industrial revolution. This finding motivated new studies of global climate changes. Although the results confirmed Arrhenius's predictions, a slight decrease in globally averaged temperatures in the 1960s and early 1970s prompted some scientists to warn the public that an ice age may be imminent! The story changed when the 1980s brought an increase in temperatures. By the late 1980s the public was paying attention to global warming, in part because 1988 happened to be an exceptional year in terms of weather phenomena: the United States experienced such a horrendously hot, dry summer that cattle had to be slaughtered for lack of grass to feed them; winds swept topsoil into dark clouds reminiscent of the Dust Bowl days of the 1930s; newspapers and television showed pictures of barges stranded in the Mississippi River, which was running dry; forest fires ravaged millions of acres in the West; in the eastern states temperatures were so unbearably high that assembly lines were shut down in some factories. The Soviet Union and China were similarly drought stricken, but torrential rains plagued parts of Africa, India, and Bangladesh, and exceptionally intense hurricane Gilbert swept across the Yucatan Peninsula. At the end of that year, the cover of *Time* magazine had a picture, not of the "Man of the Year," but of planet Earth, a planet in peril.

Several events in the 1980s contributed to heightened public awareness of global climate changes. First there was intense El Niño of 1982, followed shortly afterward by the alarming announcement that the ozone layer, which protects us from dangerous ultraviolet rays in sunlight, had developed a gaping hole over Antarctica. Finally there was the horrendously hot, dry summer of 1988, which some scientists attributed to the rise in the atmospheric concentration of greenhouse gases. (Many people confuse global warming with the ozone hole and are under the impression that they are one and the same thing.) Some scholars argue that the political upheavals that accompanied the end of the Cold War and that prompted extensive discussions of global political changes, and of a "global market economy," were additional factors that contributed to heightened interest in global changes in general,[3] global climate changes in particular. (In some European countries, politicians became concerned about global warming when they learned that the associated climate changes could result in third-world people fleeing the tropics and seeking refuge in higher, cooler latitudes.)

The political changes of the 1980s directly affected scientific research when the sponsors of that research started changing their reasons for funding scientific projects. During the Cold War the U.S. government seemed to be very generous in funding research with no immediate or even obvious long-term benefit to the state, in the fields of oceanography and meteorology for instance. One possibility is that the government did this because it was interested in "maintaining a *labor force* of skilled scientists available for consultation on policy issues." "A pool of scientists actively engaged in research will have, at least in theory, highly developed investigative and problem-solving skills that can be called upon to address immediately pressing problems."[4] Such motives did not interfere with the main goal of the scientists studying El Niño, namely, an understanding of a fascinating natural phenomenon. Producing useful results was of secondary importance until, in 1982, a devastating El Niño caught everyone by surprise. Shortly afterward the end of the Cold War brought increasing pressure on scientists to demonstrate that their results are of practical value. By emphasizing the damage El Niño can inflict, scientists could use that phenomenon as an effective means for increasing the resources available for studying the atmosphere and oceans. The scientists had a splendid opportunity to demonstrate the usefulness of their activities when, in May 1997, instruments maintained in the tropical Pacific showed that yet another intense El Niño was developing. It was very likely that large parts of the globe would be affected. The scientists therefore called press conferences to alert the public, and also government officials in the United States and in other countries, of impending disasters. Because of their efforts, the subsequent development of El Niño, over a period of many months, received far more coverage in the press and on television than did the comparable event of 1982. Everybody became familiar with El Niño in 1997. In Great Britain, a country not significantly affected by this phenomenon, only the Spice Girls, a singing group, had more column inches devoted to them in newspapers and tabloids than did El Niño. The scientists succeeded far beyond expectations in drawing attention to this phenomenon, in part because of intriguingly parallel developments in the world of human affairs.

The financial world was in turmoil in 1997. The global market economy, which everyone trumpeted after the end of the Cold War, seemed to have gone awry. The problems were most serious in Korea and Indonesia—those countries were said to be experiencing an economic meltdown—but it soon was evident that quivers in Asia cause shivers on Wall Street, that Argentina shakes when Mexico quakes, and that trouble with the ruble is worrisome to everyone. El Niño, by making an appearance during these global economic upheavals, reinforced the message that we are all interconnected. He did this by demonstrating how a warming of the waters of the remote eastern tropical Pacific Ocean can affect all parts of the planet, bringing a lack of rain to southeastern Asia but an excess to California and Kenya, mild winters to Canada, and strange weather to much of the globe. Even his march from west to east across the Pacific seemed to mirror that of the financial crisis. In June 1997 El Niño brought unusually dry conditions over southeastern Asia, contributing to vast forest fires that blanketed cities such as Djakarta and Singapore with thick smoke, so thick that it was a major factor in the crash of a commercial airplane. Over the course of the next several months the warmest waters of the equatorial Pacific, which are usually west of the date line, surged eastward until they reached South America, whereafter the warm waters spread pole-ward along the coast, to California and Chile. By that time news of El Niño, and of the crisis in the "global market economy," had brought everyone to the realization that, on this planet, "no man is an island."

FOUR

A Matchmaker

Opposites attract.
—Anonymous

The need to take care of a child, El Niño, has led to the marriage of a very odd couple, oceanography and meteorology. So disparate are the cultures and communities of these two disciplines that a successful union seemed improbable, but it nonetheless is thriving. Does this mean that oceanographers, those romantic, intrepid explorers of the watery part of the globe, are not as independent and adventurous as they used to be? How will they change now that they are wedded to those assiduous, diligent meteorologists who provide us with weather forecasts each and every day?

We have an ambivalent relationship with meteorologists. We pay close attention to the information they provide — it is readily available in newspapers, on radio, and on television — because timely alerts of imminent storms, and of the whereabouts of tornadoes and hurricanes, are invaluable to us. However, we remain steadfastly skeptical of those forecasts. Everyone can gleefully recall a recent occasion when, instead of the predicted sunny skies, it rained torrentially. If the weather forecasts are as inaccurate as many people seem to believe, why do those forecasts generate such enormous interest? Could it be that the predictions, though useful much of the time, occasionally provide reassuring and much needed confirmation of the romantic belief that nature is capricious and unpredictable? Does weather give us opportunities to have our cake and eat it?

Oceanographers serve us very differently. They provide us with vicarious thrills when we watch television documentaries of daring expeditions to explore the mysterious oceans. The evocative names of their ships — *Calypso, Atlantis, Challenger, Meteor* — add to the aura of romance and excitement. Documentaries about the oceans dwell at length on its beguiling inhabitants, colorful fish, playful dolphins, and vicious sharks. Numerous color photographs of these denizens of the seas adorn college textbooks for oceanography courses. Similar pictures of the birds, bees, and butterflies that swim in the atmosphere are absent from books that introduce the reader to meteorology. When the latter pictures appear on television, they are understood to be from the worlds of the ornithologist, apiarist, and lepidopterist, not that of the meteorologist. We regard atmospheric scientists as practical specialists, oceanographers as romantic generalists.

The different ways in which we view meteorologists and oceanographers reflect the different relationships we have with the atmosphere and ocean. So involved are we with changes in atmospheric conditions, with weather, that it is a topic of daily conversation. In different parts of the globe people take pride in the peculiarities of "their" weather and the way it has shaped their cultures. We live in the atmosphere — without it we survive for a few minutes at most — but we regard the ocean as another world, one whose mysteries fascinate us and whose hazards frighten us. Sometimes the ocean moves us to indulge in unusual behavior. For example, when on the beach, we are inclined to lose our inhibitions and to shed our clothes. (We usually refrain from such behavior in most other public places.) Sometimes we flock to the shore, not for recreation, but for meditation:

Posted like silent sentinels all around the town, stand thousands upon thousands of mortal men fixed in ocean reveries. Some leaning against the spiles; some seated upon the pier-heads; some looking over the bulwarks of ships from China; some high aloft in the rigging, as if striving to get a still better seaward peep. But these are all landsmen; of weekdays pent up in lath and plaster — tied to counters, nailed to benches, clinched to desks. How then is this? Are the green fields gone? What do they here?

Herman Melville, Moby-Dick

"Mortal men" still look on the ocean with reverence, awed by its infinity, its mystery. They used to think that, below its restless surface, the ancient ocean was immutable, but over the past several decades they have discovered that the oceans are in a constant state of flux. Over millions of years, the continual drifting of the continents causes some ocean basins to disappear and new ones to appear. (There was no Atlantic Ocean some 250 million years ago; since then the Pacific has been shrinking, the Atlantic expanding.) Over several centuries there are gradual changes in the circulation (or conveyor belt) that carries cold, saline water around the globe, from the surface into the deep ocean in high latitudes, followed by very slow motion across the ocean floor into the various basins, and finally ascent back to the surface. Motion is far more rapid in the upper ocean, especially in the swift Gulf Stream and Kuroshio Current, salient features of the basinwide gyres that the winds maintain. The low latitudes too have intense currents, capable of dramatic changes, even reversals, over periods as short as a few weeks or months. This happens seasonally in the Indian Ocean, in response to the reversal of the monsoon winds, and in the Pacific during El Niño. The ocean is dynamic, not static. It constantly interacts with the atmosphere to determine our weather and climate. Because of this new perspective we have started to look at the oceans with the eyes of meteorologists; we try to discern continually evolving patterns in order to anticipate what will happen next. This development is bringing a radical change to the field of oceanography, a change from expeditions in solitary vessels to automated arrays of instruments that monitor the tropical Pacific. This change, depicted in figure 4.1, seems to imply that the spirit of adventure (symbolized by the HMS *Challenger*) is being curbed in order to honor a firm commitment: keeping a watchful eye on El Niño by maintaining the TOGA[1] array of instruments.

Traditionally, oceanographers are explorers who do science for

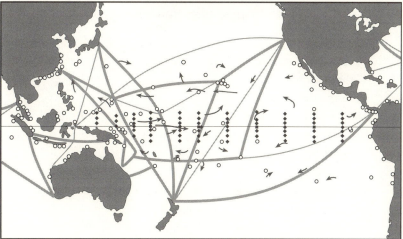

Figure 4.1. HMS *Challenger*, the corvette that launched the science of ocean-ography when it circumnavigated the globe between 1872 and 1876, and the TOGA array of instruments, which currently monitors oceanic conditions in the tropical Pacific Ocean.[1]

the sake of science. Atmospheric scientists, on the other hand, have an obligation to pursue science for the sake of being use-ful—producing a daily forecast. For the union of oceanographers and meteorologists to flourish, the two parties have to find a bal-ance between these two intertwined goals of science. This is no trivial matter, but there are good reasons for optimism. Ocean-ographers, by maintaining the TOGA array, are demonstrating their commitment to being useful. Meteorologists appreciate the importance of understanding atmospheric phenomena — such knowledge has contributed significantly to the impressive advances in weather prediction. The two parties are therefore off to a good start. Of the two, meteorology is the more experienced. Idealistic oceanography is acquiring a seasoned partner, wise in the ways of the world. Exactly what experiences have the meteorologists had?

For a long time we regarded the prediction of the weather as a form of witchcraft. For example, the character Henchard in Thomas Hardy's novel *The Mayor of Casterbridge* has the following exchange with "a man of curious repute as a forecaster or weather-prophet":

"Now, for instance, can ye charm away warts?"

"Without trouble."

"Cure the evil?"

"That I've done — with consideration — if they will wear the toad-bag by night as well as by day."

"Forecast the weather?"

"With labour and time."

"Then take this," said Henchard. "'Tis a crown-piece. Now, what is the harvest fortnight to be? When can I know?"

"I've worked it out already, and you can know at once. (The fact was that five farmers had already been there on the same errand from different parts of the country.) "By the sun, moon, and stars, by the clouds, the winds, the trees, and grass, the candle-flame and swallows, the smell of the herbs; likewise by the cat's eyes, the ravens, the leeches, the spiders, and the dung-mixen, the last fortnight in August will be — rain and tempest."

"You are not certain, of course?"

"As one can be in a world where all's unsure."

Thomas Hardy, *Mayor of Casterbridge*

Presumably the weather prophet made his forecasts on the basis of patterns he discerned in the clouds, the winds, the smell of herbs, and so on. Those whose livelihood depends on the weather, sailors and farmers for example, are often very knowledgeable about patterns that foretell imminent changes in the weather. "Red skies at night, shepherds delight" is a reminder that a break in the clouds in the west, one that permits the sunset to be red, is often an indicator of fair weather the next day. This adage is frequently correct because storms generally move from west to east (outside the tropics). Such movements can readily be tracked on maps of atmospheric conditions over a large region. The preparation of such maps, a highly organized activity that requires the collection, in a central location, of data gathered over that region, became feasible after the invention of the telegraph in 1843. That development prompted many nations to establish national meteorological services and to expand significantly the network of instruments that monitor atmospheric conditions. During the subsequent decades the availability of more and more measurements led to an improved understanding of weather phenomena and to advances in weather prediction.

After World War II, the development of the electronic computer led to a radically different method for predicting the weather. Rather than forecasts by means of extrapolations based on how weather patterns have evolved over the past few days, the computer makes it possible to solve the equations (the laws) that govern atmospheric motion and thus to determine mathematically how the patterns will evolve. The complexity of weather proved too much for the first, very modest computers of the 1950s so that they had very little success in predicting changes in the weather several days in advance. (Weather patterns tend to persist for several days, so that a person who predicts that the weather tomorrow will be the same as it is today is often likely to be right. For a long time, such a forecast was more accurate than that produced by a computer.) Advances in the design of computers, and in our ability to observe and explain atmospheric phenomena — in the 1960s it became possible to make measurements by means of artificial satellites — led to significant improvements in weather prediction. The public gradually came to accept weather forecasting as a scientific enterprise, albeit one with such limited accuracy as to provide comic relief from more serious news — in the 1960s weathermen on television were regarded as entertainers rather than providers of reliable information. By the 1970s computer predictions started to rival those of traditional weather forecasters who made extrapolations on the basis of changes observed over the past few days. Subsequent progress has been such that, today, weather is important news that receives ample attention in the press; there are even television channels devoted entirely to the weather, twenty-four hours a day, seven days a week. As with sports and politics, major events such as severe storms, floods, and hurricanes are covered on-site (by OCMs, or on-camera meteorologists), while experts in studios provide speculative, entertaining analyses. (Droughts, which develop more gradually and which are less telegenic, receive less attention.) Some people still regard weather prediction as an augury, and some still regard damaging storms as punishment for our wicked ways, but even those people are now joining everyone else in paying more and more attention to the wealth of information about the weather that is available in the press and on television. To scientists, the seemingly onerous task of predicting the weather each day brings enormous resources to the field of meteorology — a global network of instruments, supercomputers to cope with the measurements and to make forecasts — and in addition gives that field cohesion by integrating the efforts of an army of instrument makers, computer engineers, observers, and theorists. By finding a balance between science for the sake of understanding natural phenomena and science for practical benefits, meteorologists have advanced their field considerably.

Will further progress permit accurate predictions of the weather on a specific day several weeks hence? The meteorologist Edward Lorenz of the Massachusetts Institute of Technology addressed this question in the 1960s. His work, which launched a new branch of science popularly referred to as the study of chaos,[2] established that there are limits to the predictability of weather. The current computer models of the atmosphere are approaching those limits, which are on the order of ten days.

We may be unable to predict the weather on a specific day two weeks hence, but that does not undermine our confident expectation, on January 1, that temperatures on July 4 will be much higher. Our expectation is based on familiarity with what has happened in the past: the increase in the intensity of sunlight from January to July invariably results in higher temperatures in July. To some degree the sunlight heats the atmosphere directly, but to an even greater degree it heats the atmosphere indirectly by first warming up the surface of the earth. The surface then warms up the atmosphere. Changes in the temperature patterns at the surface of the earth can be unimportant to changes in the weather from one day to the next, but they play a central role in climate changes over many months. If so, then even models that are unable to anticipate the weather more than a few days hence should be capable of simulating changes in Earth's climatic patterns over many months, provided that changing temperature patterns at the surface of the earth are specified. A crucial test for the models is to reproduce the effects of El Niño, a warming of the eastern tropical Pacific, on the atmosphere. The models are indeed capable of simulating those effects, provided that the sea surface temperature changes are specified. This remarkable result implies that, to make long-term forecasts, atmospheric scientists need to join forces with oceanographers.

The realization that the marriage of the atmospheric and oceanic sciences could be beneficial to both parties is a very recent development and came after many unsuccessful attempts to anticipate long-term changes in climate. Those efforts were already under way in the nineteenth century, for example in India, where a failure of the monsoons used to be associated with horrendous famines. To cope with such disasters, the authorities considered forecasts of the monsoons to be of vital importance. After the catastrophic famine of 1899 Gilbert Walker, a versatile young mathematician from Cambridge University, was assigned the task of developing methods for the prediction of the monsoons.[3] At the time there was no scientific understanding of the phenomenon that could guide Walker's efforts. He therefore followed the advice of a prominent contemporary, Norman Lockyer, the editor of science journal *Nature*:

> Surely in meteorology, as in astronomy, the thing to hunt down is a cycle, and if that is not to be found in the temperate zone, then go to the frigid zones and look for it, or the torrid zones and look for it, and if found, then above all things, and in whatever manner, lay hold of it, study it, record it, and see what it means.[4]

Walker started a search for patterns by means of statistical analyses of meteorological data from around the globe, looking for "all sorts of curious coincidences." (A visitor to his office in Simla described "rows and rows of pigeonholes" in which he stored his results).[5] That was how Walker uncovered the Southern Oscillation, a coherent, large-scale fluctuation that involves the movement of air masses back and forth across the tropical Pacific and Indian Oceans — "when pressure is high in the Pacific Ocean it tends to be low in the Indian Ocean from Africa to Australia," and vice versa.[6] When the warm, moist air over the western tropical Pacific surges eastward, toward South America, India tends to have poor monsoons. When the moist air returns westward, that subcontinent usually has plentiful rainfall. (A complete cycle lasts on the order of four years, but the oscillation can be very irregular.) Although Walker's documentation of the Southern Oscillation was an important scientific contribution, he could neither explain that phenomenon nor convert his results into a scheme that would predict the monsoons, in part because he lacked a vital clue. His work fell into oblivion until that clue became available in 1957, a very auspicious year in our affair with El Niño.

Up to 1957 we regarded El Niño as a regional phenomenon confined to the shores of Ecuador and Peru. We gained that impression from the data gathered over the centuries by vessels that sailed across the Pacific. From these oceanographic data we learned that, along the equator, the Pacific is generally warm in the west, near the international date line, and cold off South America except during El Niño's occasional visits, when there is a temporary warming of the waters in the east. We remained ignorant of the spatial extent of that warming because the measurements acquired by the solitary research and commercial vessels plowing the seas provided very little information about temporary or periodic changes in oceanic conditions. Because there were no data that indicated otherwise, we came to regard El Niño as a sporadic, regional climate curiosity along the western shores of South America.

We changed our minds concerning El Niño when that phenomenon appeared in 1957 because, that year, scientists happened to be conducting the International Geophysical Year (IGY) of enhanced measurements of our planet, a huge, international research effort that involved sixty-seven nations.[7] The measurements acquired during IGY revealed that the warm waters associated with El Niño extend thousands of miles westward from the shores of Peru and Ecuador, across the entire ocean basin. When this surprising result came to the attention of Professor Jacob Bjerknes, the protagonist in the story of our affair with El Niño, he immediately realized that a change over such a vast area must have a profound influence on the atmosphere.[8] Bjerknes realized that there must be a link between El Niño and Walker's almost forgotten Southern Oscillation.

The demands of an empire had taken Gilbert Walker from Great Britain to India, where he discovered the Southern Oscillation. The horrors of World War II forced the Norwegian meteorologist Jacob Bjerknes to remain in the United States when he happened to be there on an extended visit in 1940. Bjerknes found himself stranded when the Germans invaded Norway. He joined the University of California in Los Angeles, where he established a training school for U.S. Air Force weather officers. In the 1950s, with support from the inter-American Tropical Tuna Commission, he started studying El Niño. Bjerknes was a distinguished, much decorated meteorologist when he turned his attention to this phenomenon. (The son of a famous Norwegian meteorologist, Bjerknes had made his mark shortly after World War I with pioneering research on the dynamics of weather in high latitudes.) When data acquired during the special measurement program of 1957 revealed that the warming associated with El Niño was not confined to the coast of Peru but extended across the entire tropical Pacific, Bjerknes immediately appreciated that, if the warming was as extensive every time El Niño visited, then an explanation for Walker's Southern Oscillation would be readily available because temperature patterns at the surface of the earth determine its climate. Over the warmest regions in low latitudes, moist air rises spontaneously into towering cumulus clouds that provide plentiful precipitation locally. Aloft, the air that is drained of its moisture flows outward from the clouds and subsides over regions with relatively low surface temperatures. Hence warm regions in the tropical Pacific have plentiful rainfall; relatively cool regions have little precipitation. The region of heavy rains and warm waters is confined to the far western Pacific before El Niño but expands eastward toward the coast of South America during that phenomenon, whereafter it contracts back westward. Walker's Southern Oscillation, the continual movement of air masses back and forth across the Pacific, is therefore a consequence of the sea surface temperature changes in the eastern tropical Pacific.

Sea surface temperature variations cause the Southern Oscillation. But what causes the changes in ocean temperatures? Why are the waters of the eastern Pacific warm during El Niño but otherwise cool? The determining factor in the tropics is not the flux of heat across the ocean surface (as is the case in higher latitudes) but the winds. Their fluctuations—the trades are relaxed during El Niño but intense otherwise—can translate into the interannual temperature changes observed in the Pacific. These changes in the winds, along with the shifts in rainfall patterns mentioned earlier, are all aspects of the Southern Oscillation. This means that the sea surface temperature changes are both the cause and the consequence of the Southern Oscillation. From this circular argument Bjerknes inferred that this phenomenon, and El Niño, are the products of interactions between the ocean and atmosphere.

El Niño of 1957 was an epiphany. Seemingly unrelated atmospheric and oceanic phenomena, previously thought to be regional, suddenly acquired grand dimensions and breathtaking unity. It became clear that El Niño is but one phase of the continual Southern Oscillation, which affects much of the globe and which arises because of interactions between the ocean and atmosphere involving the entire tropical Pacific. Bjerknes arrived at these remarkable insights on the basis of very scant data sets and qualitative reasoning. To place his results on a firm footing would require considerable resources for detailed observational and theoretical studies of the ocean and atmosphere, all with the goal of understanding and predicting El Niño and, hopefully, the monsoons.

Bjerknes's seminal work on El Niño and the Southern Oscillation provided a blueprint for melding into a comprehensive program the diverse studies required to understand those phenomena better. By the 1980s many of Bjerknes's conjectures had been confirmed, but that work was motivated, not so much by Bjerknes's results, which were published in the 1960s, as by political developments that accompanied the launching of *Sputnik* in that auspicious year, 1957.

Scientists in the United States were the first to propose putting a satellite into orbit around Earth as a contribution to the International Geophysical Year, but they encountered difficulties and delays. They, and everyone in the West, were astonished and dismayed when, in October 1957, the former Soviet Union launched *Sputnik*, the first artificial satellite to orbit Earth. Could the United States be lagging behind the Soviet Union in science? Such concerns resulted in a bonanza for science in general. The atmospheric and especially the oceanic sciences benefited for an additional reason: an artificial satellite is a marvelous platform from which to observe our planet and its atmosphere. (A polar-orbiting satellite circles Earth every ninety minutes; geostationary satellites remain vertically above a fixed point on the equator.) Instruments on such a platform can readily "see" through the atmosphere to the surface of the earth but are unable to "see" below the surface of the ocean. The military importance of the oceans therefore increased after *Sputnik*; submarines carrying nuclear weapons can effectively be hidden below the sea surface. The need to know more about oceanic conditions, especially the variability of those conditions, led to a huge increase in the resources available to oceanographers.[9]

Traditionally, oceanographers made measurements from isolated vessels on relatively brief expeditions to different parts of the globe. They were thus able to study phenomena that could be considered "permanent," such as oceanic currents like the Gulf Stream, but they had essentially no information about the variability of those currents. Oceanographers were still discovering major features of the oceanic circulation as recently as 1953, when they accidentally came across the Equatorial Undercurrent. This remarkable subsurface eastward jet, which flows counter to the prevailing surface winds, is comparable in intensity and transport to the Gulf Stream and amounts to a narrow ribbon that precisely marks the location of the equator across the full width of the vast Pacific Ocean. An explanation for this current was still being debated in 1957, as is evident from the announcement of a colloquium in 1959, shown in figure 4.2.

Exciting discoveries in oceanography came at an even faster pace after *Sputnik* because of an important by-product of that satellite: a large increase in the number of international research programs to study the oceans and the atmosphere. The predecessors of the International Geophysical Year of 1957 — the International Polar Years of 1882 and 1932 — were fifty years apart, but after 1957 there was a huge increase in the frequency of similar programs involving scientists from several institutions and nations, including the United States and the Soviet Union. These collaborative efforts to study various parts of the globe presumably helped to ease Cold War tensions. When President Lyndon Johnson proposed that the 1970s be the International Decade of Ocean Exploration (IDOE), he quoted Longfellow ("the dim, dark sea — that divides — yet unites mankind") and expressed the hope that exploration of the oceans would promote peaceful cooperation between nations.[10] Similar sentiments motivated atmospheric scientists to launch the massive Global Atmospheric Research Program (GARP), in collaboration with oceanographers, under the auspices of the World Meteorological Organization, an agency of the United Nations.

The large, coordinated programs enabled oceanographers to expand the focus of their studies from the "climate" of the ocean

NOTICE

at the M.I.T. Faculty Club.

3:45 PM MAY 28, 1959 FRIDAY

A COMPETITION of

THEORIES,

of the

EQUATORIAL UNDERCURRENT

ALIAS "THE CROMWELL CURRENT"

FEATURING :

DR GEORGE VERONIS, PROF. J. G. CHARNEY &
HENRY STOMMEL, ESQ. WITH THREE DIFFERENT
MODELS EACH IN ONLY FIFTEEN MINUTES
and
PROFESSOR W. V. R. MALKUS PRESENTING THREE
ANTIMODELS, EACH IN FIVE MINUTES
and
A SURPRISE APPEARANCE OF PROFESSOR
R. S. ARTHUR WITH A DISCUSSION of SOME
RELATED FACTS OBSERVED IN THE PACIFIC

BREATHLESS PERFORMANCE !

ADMISSION FREE

Figure 4.2. A colloquium notice prepared by Henry Stommel.

to include oceanic "weather" too. This new interest put oceanographers in an unusual situation, very different from the one
experienced by atmospheric scientists. In meteorology, data sets
that describe changes in atmospheric conditions have been gathered for a long time because interest in the weather has always
prompted enthusiastic amateurs, spread over huge continental
areas, to make routine measurements. (Several of the early presi-
dents of the United States maintained weather logs. Jefferson did
so for fifty years, starting in 1777.) The invention of the telegraph
in 1843 led to such an expansion of the meteorological networks
that Walker in India at the beginning of the twentieth century,
and the young Jacob Bjerknes in Norway at the time of World
War I, both had access to extensive data sets to test and guide
their ideas concerning climate and weather. Oceanographers, by
contrast, had almost no information about the variability of currents when, in the 1960s, they started to plan large programs
to study that variability. Limited resources and logistical constraints — the time a research vessel can stay at sea, or an instrument at sea can remain unattended, is short — meant that only
tiny parts of the vast oceans could be observed in a manner suit-
able for the study of time-dependent motion. On which regions
should there be a focus? How should the instruments be arrayed
in those regions?

Because of questions such as these, the ranks of the oceanographers increasingly included theorists interested mainly in mathematical descriptions of oceanic motion. To them, the ocean and
atmosphere have striking parallels because they are both stratified
fluids on a rotating sphere, the solid earth. (At the Massachusetts
Institute of Technology some students referred to the oceans as
"wet air.") At first, these new members of the oceanographic
community — they did not participate seriously in the acquisition
of measurements — produced highly idealized, mathematical models
of the oceans that yielded results with a "dream-like" quality,
results that seemed to have little relevance to reality. However, it
soon became clear that their ideas concerning waves and currents
near the equator are relevant to the appearance of unusually warm
water in the eastern equatorial Pacific during El Niño. Further-

more, the scientists realized that they could considerably enhance their understanding of certain oceanic processes (and of El Niño) by stepping beyond the confines of the Pacific, by exploring related phenomena in all three tropical oceans. The similarities and differences between the three oceans are such that they in effect are three different laboratories for studying how different factors influence the oceanic response to changes in the winds. Each of the three basins has a complex maze of warm and cold, eastward and westward, surface and subsurface currents, but there are significant differences between them. The Pacific and Atlantic Oceans have such vastly different dimensions that they have different manifestations of the seasonal cycle and of interannual fluctuations. The Atlantic and Indian Oceans are comparable in size, but whereas smoothly varying easterly trade winds prevail over the tropical Atlantic, abruptly changing monsoons force the Indian Ocean. By demonstrating that seemingly different phenomena in the three oceans are all in accord with the same basic principles, and can be simulated realistically by means of models that incorporate those principles, theorists established confidence in their models.

Progress was so rapid that, by the time El Niño visited in 1982, scientists were able to document, in considerable detail, the evolution of El Niño across the Pacific. Furthermore, by 1982 they had developed models of the ocean — the counterparts of the atmospheric models that are used for weather prediction — to explain and simulate that event. The key result that emerged from these studies is that the warming of the eastern tropical Pacific during El Niño is part of the response of the ocean to changes in the winds over the entire ocean, especially the far western equatorial Pacific. The computer models of the ocean proved capable of realistic simulations of that response, including the temperature changes of the eastern Pacific, provided that the changes in the winds were specified. Up to this time, oceanographers had been focusing on simply understanding oceanic phenomena rather than providing useful information. Hence the detailed descriptions and simulations of El Niño of 1982 became available only after the phenomenon had occurred. So devastating was that event, so much publicity did it receive, that it was obvious that, in future, the public would have to be alerted to El Niño's visits in a timely manner.

El Niño of 1982 was a landmark event that led to several developments. One was the deployment of the TOGA array of instruments to monitor the tropical Pacific. The advances in technology that permitted this step are still affecting all aspects of oceanography. Consider a matter as basic as determining the position of a ship at sea. Today, the amazing Global Positioning System (which depends on the presence of several special satellites in the sky) enables us to determine our position anywhere on the globe, under any atmospheric conditions, with astonishing accuracy. This facility has been available for little more than a decade now. Previously, oceanographers determined their position at sea by observing the stars. Satellites have also increased the accessibility of measurements made at sea enormously. To appreciate their impact, consider that the data collected by the HMS *Challenger* in the nineteenth century, although a relatively small amount, appeared over a period of many years after the voyage had ended and were finally published in fifty large volumes — 29,500 pages in all. During the next century, it was standard practice for oceanographers to spend months and years analyzing the data they had collected at sea before making that data available to the larger community. Today, the torrent of measurements that flows from the TOGA array of instruments in the Pacific is available immediately to everyone. As in the case of atmospheric data, computer models of the oceans now assimilate the data from the TOGA array and routinely make available maps of conditions in the tropical Pacific, the counterpart of daily weather maps. (The measurements are made at points that are far apart; the models in effect interpolate between those points.)

Another consequence of El Niño of 1982 was wide recognition of something Bjerknes had already realized in the 1960s, namely, that the phenomenon is the product of interactions between the ocean and atmosphere. This realization has important and also disconcerting implications. We are forced to acknowledge that nature often disregards the distinct categories into which we, in our efforts to cope with a complex world, divide nature's myriad interacting phenomena. On some university campuses, oceanographers and meteorologists inhabit separate institutions or departments and have different interests. Oceanographers traditionally study the effect of the winds on the oceans, but the causes of those winds, and the reasons for their fluctuations, are the concern of meteorologists. Atmospheric scientists study the response of the atmosphere to changes in sea surface temperatures but leave it to their oceanographic colleagues to explain why the temperatures change. For a long time, such a division of labor made sense. For example, when the University of California started establishing campuses in several cities, it seemed logical to avoid duplication of disciplines with small numbers of practitioners. Hence only the campus in Los Angeles was allocated a department of atmospheric sciences. Oceanography was assigned to San Diego, more than a hundred miles away. The concentration, in one place, of all the activities in a relatively small field seemed logical, until El Niño made us realize that our seemingly reasonable categories and compartments have arbitrary aspects. Today the University of California has oceanographers in Los Angeles, meteorologists in San Diego.

The recognition that El Niño is dependent on interactions between the ocean and atmosphere is so recent and sudden that many people still have difficulty coping with this reality. That is why we still encounter, in newspapers, magazines, and even in scientific journals, articles that explain and discuss El Niño either in strictly atmospheric or in strictly oceanic terms. Some people attribute this phenomenon to changes in tropical sea surface temperatures (without asking why the temperatures change) while others believe that it is a consequence of changes in the winds over the tropics (without asking why the winds change). In scholarly journals, similarly incomplete perspectives are expressed in more technical terms when El Niño is attributed to oceanic Kelvin waves, for example. (Those waves are strictly oceanic phenomena; to assign them the principal role in the development of El Niño amounts to a denial that ocean-atmosphere interactions are at the heart of the matter.) To remedy such narrow perspectives, oceanography students are now expected to be familiar with the atmospheric sciences too. In 1942 the newly designed curriculum for students at the Scripps Institution of Oceanography (in San Diego) covered four topics: physical, chemical, geological, and biological oceanography.[12] Conspicuously absent was explicit mention of the atmospheric sciences. Today the importance of meteorology is more explicit in curricula, but unfortunately there is no longer an insistence that students need to know about all facets of the oceans. Hence some researchers are specialists in the motion of the ocean and atmosphere but take little interest in the biology or geology of the oceans. El Niño has enriched everybody by joining together two previously independent disciplines. We need to take care not to exclude anyone from this extended family, because we face serious problems that require the close collaboration of all earth scientists.

The climate of this planet, and hence its habitability, depend on interactions between its different components: the atmosphere, oceans, solid earth, and biosphere (the assemblage of all life-forms). It follows that, to anticipate climate changes—to predict how global warming will affect different parts of the globe, for example—we have to integrate the traditionally separate fields of geology, zoology, botany, meteorology, and oceanography. Natural historians recognized already in the early nineteenth century that the subdisciplines of the geosciences are interdependent, that the surprising fossils that can be found in the rocks are closely linked to the dramatically different climates Earth has experienced in the past. The renowned geographer Alexander von Humboldt (1769–1859) wrote that "the perception of the chain of connections by which all natural forces are linked together and made mutually dependent on one another, exalts our minds." Since then we have learned much about the continually changing conditions on Earth during the unimaginably vast spans of time in the geological record. For example, we now know that swelteringly hot conditions 100 million years ago, when dinosaurs roamed the continents, were followed by a gradual cooling of the planet that led to recurrent Ice Ages over the past 3 million years. The price of those scientific advances was increasing specialization, the splintering of the earth sciences into separate sub-disciplines. In the 1960s, confirmation of the hypothesis of continental drift was a major step toward integration of the earth sciences. This unifying idea explains diverse phenomena such as the complementary shapes of continents and the distribution of earthquakes, volcanoes, and mountain ranges. The next step toward the fulfillment of von Humboldt's vision, now under way, has as its goal not only detailed descriptions of conditions in the past but also explanations for those conditions. We know where the continents were located at different times in the past, when different mountain belts were formed, and which plant and animal species were dominant; now we propose to explain why those different conditions were associated with different climates. Why was the Cretaceous period so extremely warm? Why did the termination of the last Ice Age commence some 20,000 years ago, introducing a pleasantly warm period during which humans could thrive? These are questions geoscientists are now addressing. Successful explanations will contribute to our ability to anticipate future climate changes — those associated with the rapid rise in the atmospheric concentration of greenhouse gases, for example.

Now that El Niño has arranged the marriage of the atmospheric and oceanic sciences, we need him to assist with a much more ambitious project: the integration of the various disciplines that constitute the earth sciences so that we can address urgent questions concerning the future habitability of our planet. Our experience thus far teaches us that success will require a balance between the two intertwined goals of science, gaining an understanding of natural phenomena and producing useful results. The argument of the physicist who explained the merits of science for the sake of science by stating that, although his research would not contribute to the defense of his country, it would make his country worth defending, is difficult to defend. At the other extreme it is not helpful to insist on immediate benefits from the public's considerable investment in science by quoting Samuel Johnson in a threatening tone: "Depend on it, sir, when a man knows he is to be hanged in a fortnight, it concentrates his mind wonderfully." Today there is enormous pressure on scientists to do research that is useful, to predict El Niño for example. Walker experienced similar pressure when he arrived in India on a mission to predict the monsoons. During the years of the Cold War, by contrast, oceanographers had considerable latitude to pursue science for the sake of science. That freedom probably contributed to their failure to anticipate the arrival of El Niño in 1982, but keep in mind that Walker was unsuccessful in his attempts to predict the monsoons even though that was his focus. The obligation to produce results of practical value can be beneficial, but it must be balanced with opportunities to pursue science for the sake of understanding natural phenomena.

Over the past few decades the need to produce useful results has transformed the study of El Niño from a "small" science, involving a few individuals requiring modest resources, into a "big" science with numerous teams of researchers organized by managers who secure huge budgets. This change has brought many benefits, but it has come at a cost that is implied in figure 4.1, which pictures the *Challenger* and the TOGA array. Whereas the name and picture of *Challenger* convey a sense of adventure and excitement, the map and the acronym TOGA (for Tropical Oceans and Global Atmosphere) spell organization and discipline. That uninspired acronym is appropriate for an impersonal grid of instrumented moorings, but it tells us nothing about the emotional challenge of braving the elements to explore the unknown. The chief scientist of a research vessel who accepts that challenge has considerable autonomy in deciding where to go, what to measure, and when to make available to others the data collected during the cruise. A participant in one of the coordinated international programs to study the ocean and atmosphere—they usually have mundane or unpronounceable names such as TOGA, CLIVAR, and FGGE—has far less autonomy and is not as romantic a figure as the traditional oceanographer. However, those programs offer rich rewards that amount to far more than larger and more comprehensive data sets. Frequent meetings to prepare plans and to synthesize results afford numerous opportunities for warm professional and personal ties across institutional and national boundaries. Those meetings are not impersonal gatherings limited to the exchange of technical information but are regular reunions of friends and families in wonderful and exotic places that happen to be the sites of oceanographic institutions — Paris, Dakar, Odessa, Kyoto, Guayaquil, La Jolla, Xindao, Honolulu, Woods Hole, Venice, Cape Town. . . .

El Niño deserves credit for melding observers and theorists into a cohesive oceanographic community and then arranging the marriage of the atmospheric and oceanic sciences. But his task is not yet over. He now has to help us make that union an enduring one. He has to help us prevent the TOGA array, whose maintenance requires discipline, commitment, and dedication, from having a stultifying effect. El Niño appears to be well aware of his mission. He reveals a new, unexpected facet each time he visits, thus reminding us that there is always more to be learned about him. To acquire a better understanding of El Niño, so that we can predict him better, will require, in addition to resources such as the rigid TOGA array, opportunities to be flexible and imaginative in exploring tentative hypotheses concerning puzzling aspects of El Niño. We are counting on whimsical, spontaneous El Niño to infuse TOGA with some of the romance of a *Challenger* expedition.

Part 2 | Our Dilemma

FIVE

Two Incompatible Cultures

Should gardeners be scientific, should cooks? Botany helps gardeners, laws of dietetics may help cooks, but excessive reliance on these sciences will lead them—and their clients—to their doom.
—Isaiah Berlin (1909–1997), "On Political Judgment"

"A gulf of mutual incomprehension—sometimes . . . hostility and dislike"—separates the "two cultures" of the arts and of the sciences. The scientist and novelist C. P. Snow generated a heated debate about the relative merits of the two cultures when he made this statement in the late 1950s.[1] He identified inadequate science education in schools as a large part of the problem. A few decades later, despite the growing numbers of people with some basic education in science, and despite the increasing attention that newspapers and television devote to science, the gulf that separates the "two cultures" seems to be as wide as ever.[2] Part of the reason is that interactions between the "two cultures" are far more complex today than they were a few decades ago. At that time many people accepted science as an objective social good and believed that difficulties arise mainly because of the complexity of technical details. Now there is greater awareness that although scientific results can contribute to the solution of social problems, they can also be misused for political ends, thus exacerbating social problems. The latter happens when the scientific information is assigned either too much or too little weight when deciding on policies. To assign the appropriate weight requires an awareness of, and a respect for, the differences between the worlds of science and of human affairs.

Unawareness of the radical differences between the two cultures can lead to unnecessary confusion. For example, on television we sometimes see an interviewer introduce, first a scientist who explains that global warming is a reality, then an economist who denies that it is a threat.[3] The interviewer concludes that the experts are in complete disagreement and that global warming is a perplexing matter. Very seldom is it mentioned that the scientist and the economist deal with entirely different aspects of the problem, that the one is not necessarily contradicting the other. The scientist has the relatively simple task of determining, on the basis of well-known laws that govern natural phenomena, how our planet will respond to changes in the composition of its atmosphere. An economist has the far more difficult challenge of reducing, to a few numbers, costs and benefits that each of us weighs differently because they reflect our personal values. The serious difficulties that scientists face arise mostly because they occupy positions on both sides of the gulf. Professionally they live in the world of science, but as citizens with social concerns similar to those of businessmen, lawyers, and pilots they also live in the world of human affairs. To what extent do their social values influence their supposedly objective scientific results? To address that question we need to explore the nature of the differences between the two worlds. The following allegory is useful for that purpose but is a gross oversimplification of the environmental problems we encounter in reality. (Less idealized, more realistic problems are discussed later in this chapter.)

An Allegory

We are spending a pleasant summer afternoon in a canoe, floating lazily down a river, when someone studying a map suddenly announces that we are drifting toward a waterfall. The reaction to this news is varied: some people panic and demand that we leave the water immediately; others welcome the prospect of a thrill and insist that we get within sight of the waterfall; a few, who had been taking naps, survey the tranquil waters, declare that there cannot possibly be a waterfall on this river, and start dozing off again. How do we avoid a calamity?

To prevent a disaster we need to address two questions: How far is the waterfall? And when should we get out of the water? The first is a scientific question, the second a political question. Whereas the first question has a straightforward answer that can readily be obtained, the second is far more difficult and has a multitude of answers, each one favored by some, opposed by others. Scientists have been so remarkably successful at solving the problems they address that their methods merit close inspection. We need to find out whether their methods are at all applicable to social problems. To that end, let us examine how scientists estimate the distance between the canoe and the waterfall. We assume that the task is assigned to a pilot in an airplane above the river.

1. The pilot starts by appealing to Euclid's principles of geometry, which are shared by everyone independent of sex, race, ethnicity, or religion, even though those principles were formulated more than two thousand years ago in ancient Greece. They have been used repeatedly by innumerable scientists performing countless experiments so that the pilot has confidence in them.

2. Euclid's principles tell us how the distance to the waterfall can be estimated on the basis of certain measurements, specifically the height of the plane above the river and the angles to the canoe and to the waterfall. Once the pilot has made those measurements, he calculates the distance to the waterfall, taking great care to apply Euclid's principles with complete consistency, without any compromises or exceptions. If the pilot is under the impression that the calculations involve the sine of an angle, and if the copilot believes that it should be the cosine, then they do not choose the tangent as a compromise. Rather, they refer to standard references—the published literature—to determine who is right. The distance to the waterfall is likely to be in feet if the pilot is British, meters if he is French or German, but that does not affect the distance; the two answers, after simple conversions from feet to meters, should be the same if both pilots are competent scientists.

3. One measure of the competence of a pilot as a scientist is his willingness to estimate the distance to the waterfall, not once, but numerous times. To minimize the possibility of errors, and to build confidence in his results, the pilot repeats his measurements and calculations several times. The first estimate of the distance between the canoe and the waterfall is made when the plane is far upstream from the canoe. The exercise is repeated when the plane is above the canoe and again when the plane is farther downstream. This skeptical attitude is the salient feature of the scientific culture. Scientists practice "organized skepticism" by checking and rechecking their result endlessly so that any result is tentative, subject to revision.

Scientists check their own results repeatedly and also insist on critical reviews of their results by peers, for reasons expressed in a famous aphorism usually attributed to Sir Isaac Newton: "If I have seen farther, it is by standing on the shoulders of giants."[4] A scientist makes a contribution by building on the achievements of his colleagues. The soundness of the edifice is therefore a matter of fundamental importance that has to be checked, preferably by experts. Einstein was passionate about this matter and responded as follows to a report in the *New York Times* of May 4, 1935, on a scientific dispute involving his work:

> It is my invariable practice to discuss scientific matters only in the appropriate forum and I deprecate advance publication of my announcement in regard to such matters in the secular press.

The most recent scientific results are necessarily tentative and hence are often a topic of debate in the scientific community. The results amount to a provisional report, not a conclusive answer. New information can lead to a modification or revision of earlier results. In the case of a problem as simple as measuring the distance to the waterfall, the pilot will give his best estimate after a few attempts and will then continue with other duties. In science, the repeated efforts never cease. The acclaim Newton received after he had apparently provided definitive answers to the mystery of the forces that maintain the planets in their orbits around the sun seems to indicate otherwise:

Nature and Nature's law lay hid in night
God said, Let Newton be! And all was light.

Alexander Pope

More than two hundred years elapsed before another physical scientist, Albert Einstein, became as renowned as Newton. Should we conclude that Newton's results were so definitive that in science, or more specifically in the field of celestial mechanics, not much happened after Newton until Einstein appeared? Not at all. After Newton, several generations of scientists, including the brilliant mathematicians Euler, Lagrange, and Laplace, did work of the utmost importance by continually refining the calculations of planetary orbits. Their goal was to minimize discrepancies between orbital measurements and calculations based on Newton's laws, by taking into account factors neglected by Newton. (Those factors included the gravitational influence, not just of the sun, but of other planets too on the motion of each planet.) When, in the 1820s, astronomers found that the orbit of the newly discovered planet Uranus had puzzling features, they concluded that, either Newton's laws were invalid at very large distances from the sun, or there must be another, as yet unobserved, planet. The latter prediction proved correct and led to the discovery of Neptune. Studies of the orbit of the planet Mercury suggested the presence of yet another unobserved planet, this time very close to the sun. It was tentatively named Vulcan, but no such planet has been found; none is believed to exist. In the case of Mercury, the discrepancies between orbital measurements and calculations based on Newton's laws could not be eliminated. Those discrepancies pointed at the limitations of Newton's laws and, in the twentieth century, proved to be a crucial check for the validity of Einstein's theory of general relativity. The persistent efforts of scientists to reduce errors in their results can contribute to new discoveries and sometimes even to the verification of revolutionary theories.

All scientific results have uncertainties, but scientists continually strive to reduce those uncertainties by adopting methods that have proven enormously successful, yielding results that have transformed our daily lives. So impressive are these accomplishments that, at various times, some philosophers have proposed that the scientific method be applied to social and moral problems too. Let us adopt such an approach to determine when we, in the canoe approaching the waterfall, should get out of the water. (For guidance we turn to the philosopher Isaiah Berlin, who has written extensively and eloquently about the disastrous consequences of adopting an approach that is too scientific to solve social problems.)[5]

The first step is to identify the principles or truths that should govern our social and moral behavior and then to apply them consistently, without contradictions, to the problem at hand. Although this is a very complex task, as is evident from the very different reactions of the people in the canoe to the news of the waterfall, let us suppose that there is an Isaac Newton of social studies, sufficiently clever and wise to divine those social and moral truths. Some people, because they are stupid or confused, will fail to appreciate that these are indeed truths and will oppose the wise man. If we are to make progress, they have to be restrained. The wise man knows how our society should conduct its affairs and should be given power to implement appropriate policies. The twentieth century provides numerous examples of such "wise men" who became tyrants, the perpetrators of unspeakable horrors.

The precepts we adopt to solve social problems should be evident and acceptable, not only to a few leaders, but to everyone. Such principles can easily be identified and can readily be formulated as laws that ensure, for all of us at all times, liberty, equality, justice. . . . Everybody is in favor of such laws, but we encounter serious difficulties when we try to follow the example of scientists who solve their problems by applying the laws of nature with complete consistency and without exceptions. The problem is that perfect liberty is not compatible with perfect equality. "If man is free to do anything he chooses, then the strong will crush the weak, the wolves will eat the sheep, and this puts an end to equality. If perfect equality is to be attained, then men must be prevented from outdistancing each other, whether in material or in intellectual or in spiritual achievement, otherwise inequalities will result."[6] Compromises are essential, whatever noble precepts we adopt to guide our moral lives. We have to protect the liberty of the weak by limiting the power of the strong. We can insist on always telling the truth, except when the truth is too painful. We can demand justice at all times, but cannot exclude mercy and compassion. We can declare that organized planning is essential, provided that we acknowledge the merit of spontaneity. No matter what truths we adopt as a basis for solving social and moral problems, we always find ourselves obliged to be flexible and to strive for compromises that accommodate conflicting points of view.[7]

Adherence to absolute truths, and belief in an ultimate solution for social problems, are delusions that have caused much misery in the past. To check the hubris of those who insist that their "objective" account of the social world is right and just, we have to acknowledge that subjective and cultural considerations can color our perception of the "facts." Is the waterfall a serious risk that requires prompt action? Each person in the canoe is likely to have a different opinion, even if everyone knows exactly how far the waterfall is. The response to an imminent El Niño, even if the scientific forecast is accurate, is a complicated matter in a poor country. El Niño will adversely affect the food supply, will contribute to the spread of diseases such as malaria, will damage roads and bridges. . . . Given very limited resources, how should priorities be assigned? What matters most to people in a specific place, at a specific time? In the case of global warming, should we take prompt action, given all the other serious problems that require immediate attention? Even if scientists are able to tell us precisely how and when the effects of global warming will become evident, different people will react differently. Those whose livelihood depends on selling oil and coal, or who live in Siberia and wish for milder winters, will give one answer. Those who live in coastal zones or islands that will disappear when sea level rises will give another answer.

Unlike a scientific problem, which generally has a single solution, a social problem usually has many solutions, with any one of those solutions satisfactory to some people but not to others. To reach a reasonable compromise that most people can accept, we benefit from familiarity with the different ideals and perspectives of different peoples and cultures. We need to determine what they have in common, what it is that human beings share. This means that, in our attempts to solve difficult social problems, we must reject relativism, the belief that the experiences of different cultures are all equally subjective and equally irrelevant to the problems our particular community happens to face.

Finding solutions to the problems that arise in human affairs requires an approach very different from the way we solve scientific problems by proceeding in a strictly consistent manner from certain basic truths. Inconsistencies are absent from the simpler world of science, which, to some people, provides an escape from the more complex world of social issues. The humanist Alan Lightman, reflecting on his earlier career as a physicist, writes as follows:

> The equations have a precision and elegance, a magnificent serenity, and indisputable rightness.
>
> I remember so often finding a sweet comfort in my equations after arguing with my wife about this or that domestic concern or fretting over some difficult decision in my life or feeling confused by a person I'd met. I miss that purity, that calm.[8]

To Salina, the prince in di Lampedusa's novel *The Leopard*, science is equivalent to a drug that can provide temporary relief from the problems of this world:

> Salina thought of a medicine recently discovered in the United States of America which could prevent suffering even during the most serious operations and produce serenity amid disaster. "Morphia" was the name given to this crude substitute for the stoicism of the ancients and for Christian fortitude. With the late King, poor man, phantom administration had taken the place of morphia; he, Salina, had a more refined recipe: astronomy. And thrusting away the memory of lost Ragattisi and precarious Argivocale, he plunged into reading the latest number of the *Journal des savants*. "Les

dernières observations de l'Observatoire de Greenwich présentent un intérêt tout particulier. . . ."[9]

Isaac Newton too commented on the divide that separates the worlds of science and of human affairs when, in 1720, he remarked that "I can calculate the motions of heavenly bodies, but not the madness of people." He had just sold, at a handsome profit, shares he owned in the South Sea Company.[10] He did this after concluding that the sharp rise in the price of those shares had no rational explanation, that it had resulted from speculative excess or, more concisely, mania. Although Newton's assessment of the South Sea Company, and his actions, were known to a large number of people — he was a prominent public figure at the time — the price of the shares nonetheless continued to rise rapidly. Then, a few months later, something very remarkable happened. Even though there was no evidence that Newton's assessment of the company had been wrong, Newton reentered the market, buying more shares at a much higher price! Shortly afterward the collapse of the company cost Newton a huge amount of money.

The worlds of science and of human affairs are profoundly different, and their respective demands are sometimes in conflict. The methods of one, applied to the other, can result in confusion, even disaster. Consider what could happen should we, in our financial dealings, adopt the "organized skepticism" that scientists practice. By questioning the soundness of a certain bank and encouraging others to do the same, we can start an entirely false rumor that causes the downfall of the bank when people hurriedly withdraw all the money they deposited there. For success, banks and many other public institutions depend on the confidence of the public. Scientific tenets such as "organized skepticism" can prove calamitous in human affairs. Conversely, a spirit of compromise, essential for the resolution of divisive social issues, cannot help with strictly scientific disputes. If one person insists that two plus two equals six, another that it equals eight, then it does not follow that these two people should agree on the number seven as a reasonable answer. Newspaper and magazine articles on scientific disagreements, about global warming for example, often give equal weight to opposing points of view, in the erroneous belief that science is a democratic activity. Scientists do not conduct referenda to decide on the validity of Newton's laws.

Uncertainties in scientific results—they are minimal in relatively simple phenomena, such as the orbits of planets around the sun, but are considerable in the case of complex environmental problems, global warming for example—complicate interactions between scientists and nonscientists enormously. Suppose that the scientists in the canoe give us different estimates for the distance to the waterfall. Do their disagreements have a scientific basis—confidence in the accuracy of measurements can vary—or do their disagreements reflect subjective biases, the different opinions of optimists and pessimists about when we should get out of the water? Such questions arise whenever scientists disagree, because they live in two worlds; in addition to being scientists they are also ordinary people with social concerns. If we were to follow Einstein's practice of discussing scientific matters only in the "appropriate forum," then scientific differences could be resolved objectively, before they influence policy. Today, however, scientists cannot exclude the "secular press" from the discussion of their disagreements, especially when the topic is as important as an imminent El Niño, or global warming. We are obliged to accept that nonscientific considerations influence scientific debates. This reality has several disquieting implications. Eagerness to persuade policy makers of the reliability of El Niño predictions, say, can undermine the commitment to skepticism that is essential for success in science. To promote the social value of their forecasts, scientists will be tempted to express confidence, not skepticism. Similar problems arise in reports on contentious issues by groups of supposedly disinterested scientists. How should we react to the few scientists who dissent from the opinions expressed by the majority? If a report is of interest strictly to scientists, then dissent is of vital importance; scientific progress depends on addressing the concerns of those who disagree. In the far more complicated case of a report intended for policy makers, dissent is often regarded as an inconvenient nuisance; we prefer that policies affecting society be based on scientific results with which everyone agrees. Science, however, does not provide conclusive answers; any scientific report is provisional, subject to changes should new results become available.

Even if a scientific result were free of uncertainty — even if we knew the distance to the waterfall precisely — deciding when to get out of the water would remain a difficult problem. Persuading people with conflicting wishes and concerns to compromise is no trivial matter. To avoid making a difficult problem even more difficult, we should take care to distinguish between its scientific and nonscientific aspects.

A More Realistic Situation

The allegory of people in a canoe drifting toward a waterfall, though useful for illustrating the profound differences between the worlds of science and of human affairs in highly idealized situations, is a gross oversimplification of the problems we face in reality. With the following modifications we begin to approach the complexity of environmental problems such as global warming. Rather than a single waterfall at a definite distance, consider rapids that grow more and more hazardous. Rather than a single canoe, consider a variety of vessels — flimsy rafts, fragile boats, sturdy ships — all tethered together so that a decision to leave the water has to be made jointly by everyone. How far is the waterfall? That straightforward scientific question, with a simple answer, no longer has a counterpart. Instead, scientists can attempt to describe how the treachery of the rapids increases as we progress downstream, a very complicated matter indeed. (Global warming is sometimes reduced to a single number — globally averaged temperatures will increase by $2°$ C a hundred years hence — but that number conveys scant information; it says next to nothing about the very different climate changes different parts of the world will experience.) Suppose that we succeed in obtaining reasonably accurate scientific information about the rapids. Then we face an even bigger problem — how to persuade people in the different vessels to reach an agreement concerning the appropriate time to leave the water. To those in flimsy rafts, even modest rapids can be catastrophic. To those in sturdy ships, some waterfalls amount merely to thrills.

This modified allegory is still a far cry from the complexity of a problem such as global warming because it fails to take into account that, although it may be dangerous to remain in the water, it is also dangerous to leave the water—the riverbanks are unknown and possibly treacherous. At present, our economy is so dependent on the burning of fossil fuels that a change in our industrial activities will be very disruptive. To insist on a reduction in atmospheric carbon dioxide levels could be costly; to insist on a rapid reduction will be very costly.[11] But doing nothing will also be expensive. To weigh costs and benefits we have to answer extremely difficult questions of the following kind.[12]

Policies to mitigate the impact of global warming will benefit, not the adults of today, but our grandchildren and great-grandchildren. To what extent should we make sacrifices for the benefit of future generations? Should we spend resources now to prevent potential problems? Or should we invest the money in financial markets and allow it to increase, making it easier to deal with the problems when they arise in the distant future? Many of us have higher standards of living than did our parents, who in turn are better off than their parents were. If our grandparents had been more selfish, would it have made much difference to us?

We need to compare the cost of doing nothing about global warming with the cost of trying to minimize its impact. Doing nothing could result in an increase in human fatalities because of the spread of diseases, for example. How do we assign dollar values to such potential disasters? For an estimate we can ask people how much they are willing to pay to avoid endangering their safety, health, and natural habitat. The rich can afford more than the poor. Does it follow that a life is of less value in a poor than in a rich country? Should we make such an assumption in estimating the cost of global warming?

Are there changes to the environment that are unacceptable to everyone, changes that we should avoid at all costs? ("Determining the efficient use of child labor in the United States was made moot by the collective decision that child labor — irrespective of potential economic benefits — is morally unacceptable.")[13]

Economists are developing very sophisticated tools (models) to help us with environmental problems that involve such questions. These are tools with which to explore the implications of different answers to the issues raised above. The economic models translate, into numbers, certain assumptions concerning ethical issues. Because these models proceed from the results of a mathematical climate model and then explore the consequences of certain assumptions concerning ethical issues, they are sometimes referred to as "Integrated Assessment Models." This is an unfortunate term because it can cause confusion when different models give different results. Suppose, for example, that one model indicates that the predicted global warming is no problem at all and that it is to our advantage to increase the rate of emission of greenhouse gases into the atmosphere over the next century. Another model, however, shows that we run serious risks if we allow the atmospheric concentration of greenhouse gases to rise above the present level. Are the differences attributable to uncertainties in the science or to different assumptions concerning our values? It will be best if the ethical assumptions are stated explicitly for each model. We all share a responsibility to decide which assumptions are acceptable, which unacceptable. Presumably we as humans have much in common despite our cultural and economic differences. With the aid of economists we need to identify a set of values that we all accept and that will help us develop "optimum" policies to cope with global warming.

Deciding on an "optimum" course, agreeing on a binding international regime to address global climate changes for example, is a daunting challenge. It was attempted in 1992 when 160 nations adopted the United Nations Framework Convention on Climate Change. One of the goals was to return carbon dioxide emissions to 1990 levels by the year 2000, but in most countries emissions continued to increase. This failure led to the Kyoto Protocol of 1997. Some economists judge that the measures proposed in Kyoto will not prove very effective; they believe that a more significant reduction in greenhouse gas concentrations can be achieved at a lower cost by allowing, for example, a more gradual transition to the more efficient use of energy. How do we determine whether a strategy is effective or ineffective?

A Pragmatic Approach

Scientists have achieved spectacular successes over the past few centuries by identifying certain principles that govern natural phenomena and then very consistently applying those principles to explain those phenomena. If we assume that this is all there is to the methods of scientists and enthusiastically adopt this approach to solve social problems, then the results could prove disastrous. It is of paramount importance to recognize that the success of science depends on far more than certain principles that are invoked with absolute consistency. That success also depends on a willingness to subject any proposed solution to tests, modifying or even abandoning a solution should it prove inadequate. This powerful method of trial and error is as applicable in the world of science as in the very different world of human affairs. Macbeth, for example, adopts it when he seems to see an object he is thinking about:

> Is this a Dagger, which I see before me,
> The Handle toward my Hand? Come, let me clutch thee:
> I have thee not, and yet I see thee still.
> Art thou not fatall Vision, sensible
> To feeling as to sight? Or art thou but
> A Dagger of the Minde, a false Creation?
>
> William Shakespeare, *Macbeth*

The data available to Macbeth, the sight of a dagger, suggests a hypothesis: the object is a dagger. He tests that theory by trying to grasp the dagger and concludes that the theory is incorrect, that he is dealing with a hallucination.

Adopting the method of trial and error — promoting adaptive programs whose evolution is determined by the results from those programs — is far wiser than implementing comprehensive programs that decree a rigid course of action to reach a grand, final solution. The trial-and-error method permits the correction of mistakes at an early stage before scarce resources have been wasted. Such a strategy can help us avoid the temptation to side-step difficult political decisions by focusing instead on the science. Consider, for example, how India learned to cope with the consequences of poor monsoon rains that cause poor harvests. Although the authorities in the nineteenth century assumed that any attempts to minimize the associated disasters — famines that killed millions of people — would require accurate forecasts of the monsoons, a method of trial and error, and certain critical political changes, led to a solution that is essentially independent of scientific information (see chapter 17). The need for scientific information can be exaggerated and can be used to divert attention from divisive political issues. We have to guard against this happening in the debate about global warming. Some experts assert that any measures to minimize future global warming will be hazardous to the economy; others argue that certain steps could be beneficial. The only way to find out is to implement a few of those measures on a trial basis. (We return to this matter in the epilogue.)

Recurrent phenomena can be of enormous assistance in our efforts to bridge the gulf between the worlds of science and of human affairs because they provide us with repeated opportunities to find solutions. That is why we cope remarkably well with the enormous climate change that occurs annually, namely, the seasonal cycle. That success depends on a minimum of scientific information, in the form of a calendar. El Niño too is a recurrent phenomenon. Although not as regular as the seasons, we do know how often he appears on the average and how he tends to affect different regions. It should be possible to put this scientific information to practical use when making long-term plans. Each El Niño is of course distinct, so predictions of the specific timing and amplitude of that phenomenon will be of great value. At present the uncertainties in such forecasts are huge, in part because scientists have observed relatively few of these phenomena. In due course scientists will become familiar with a greater variety of El Niño episodes, and the uncertainties in the forecasts will decrease.

The most valuable contribution scientists can make toward our efforts to cope with environmental problems is to reduce the uncertainties in scientific results. To assist them in doing so, and to help them organize their activities, everyone, and especially the sponsors of scientific research, should be aware of several odd aspects of such research, the topic of the next chapter.

SIX

"Small" Science versus "Big" Science

One of the diseases of this age is the multiplicity of books; they doth so overcharge the world that it is not able to digest the abundance of idle matter that is everyday hatched and brought forth into the world.
—Barnaby Rich (1540–1617)

Since 1980, the number of scientific articles devoted to El Niño in scholarly journals, and the number of people writing those articles, have doubled every five years approximately. If we were to maintain this explosive growth, then fairly soon all scientists would be writing articles about El Niño and, not long afterward, every man, woman, and child on Earth would be doing the same. Today already, students of El Niño are complaining that they cannot keep pace with the flow of information. Will matters be getting worse rapidly? Perusal of the articles on El Niño suggests that a crisis may not be imminent. Only a fraction of the articles being published at present merit attention; most are of little interest. It seems as if little remains to be learned about El Niño, but in reality the phenomenon still has many puzzling features. Is there an efficient route to the solutions?

The rapid growth of science often makes its older practitioners nostalgic for the simpler days of their youth. Those who started studying El Niño in the 1960s wistfully recall a time when it was possible to stay abreast of rapid, exciting developments by making a few telephone calls each week and by regularly attending a relatively small number of meetings where all the participants—

on the order of thirty — knew each other. Small groups of scientists would plan research programs to explore one or another aspect of the atmosphere and oceans and, after persuading funding agencies to provide resources, would manage those programs themselves. The situation is very different today. The study of El Niño used to be "small science" but has become "big science." Professional science managers, because they secure considerable funds to support the enormous number of people who now study this phenomenon, have acquired a large role in organizing and directing the activities of the scientists, constantly reminding them of the need to produce useful results, of the need to improve predictions of El Niño for example. Scientists now find themselves straddling the very different worlds of science and of policy, two worlds that often are in conflict. When they are in Washington, D.C., for example, seeking funds for research, they express great confidence in the recent scientific results on which they intend to build, given the necessary resources of course. Among themselves, scientists sing a different tune; they practice "organized skepticism" and always question a colleague's intriguing new results, because they realize that scientific progress depends on a skeptical attitude. Such inconsistent behavior impedes progress and can contribute to confusion and misunderstandings unless everyone is fully aware of the radical differences between the worlds of science and of human affairs, a topic explored in chapter 5. To minimize misunderstandings, everyone should also have some familiarity with the history of science. Of special interest are studies that turn the methods of science on science itself. Those studies search for patterns in the data sets that describe how many scientific papers are written each year, who the authors of those papers are, and how often each paper is cited in other papers.

While he was at Raffles College (now the University of Singapore), Derek de Solla Price, the author of the seminal book *Little Science, Big Science*, found himself in temporary custody of a complete set of the *Philosophical Transactions of the Royal Society of London*. When he set the volumes (for the years 1662 to 1930) in ten-year piles against the wall of a room, he noted that the increase in the height of the piles, from one decade to the next, described "a fine exponential curve." The piles provided vivid visual evidence that science had been growing at a rapid rate over a prolonged period. Confirmation of this result is now available in analyses of a variety of indicators of different aspects of science: the number of scientific papers published each year, of scientists writing those papers, of the journals in which the papers appear, of the people receiving Ph.D. degrees in science, and so on. All these indicators consistently show that, over many decades since the early 1700s, science has grown exponentially—at compound interest. Its rate of growth is proportional to its size, so that the bigger science becomes, the faster it grows. This growth is astonishingly rapid: the number of scientists, and of scientific articles, doubles every fifteen years approximately. Over a fifty-year period the number of scientists increases by almost a factor of ten. Thus the thousand scientists in the United States in the year 1800 had increased to ten thousand by 1850, to a hundred thousand by 1900, and to a million by 1950. A recent study by John Suppe at Princeton University indicates that since the 1970s the rate of growth of the total number of scientists has decreased somewhat but that some subdisciplines continue to grow very rapidly.[2] The study of El Niño appears to be such a subdiscipline.

To appreciate the significance of the rapid growth of science, we have to compare it to the far slower rate at which the human population grows: a doubling time of forty-five years approximately for humans, fifteen for scientists. Because of the relatively slow growth of the human population, a large fraction of the people who have ever lived are dead today. In the scientific world, matters are very different. A scientist who is active for forty-five years—the length of a typical career—finds that the size of his community doubles during the first fifteen of those years, quadruples over the first thirty years, and grows eightfold over the forty-five years. It follows that of every eight scientists who have ever lived, about seven are alive today. The number of scientists active at any time represents a very large fraction of all the scientists who have lived up to that time. As a result, science has always had a sense of immediacy, of being modern and contemporaneous. Whereas the history of humanity is mostly about dead people, the history of science is mostly about the present. That has always been the case, so that every generation of scientists has faced similar problems associated with the rapid growth of science. Complaints about too many articles being written about El Niño today are reminiscent of the statement made by Barnaby Rich (see the epigraph at the beginning of this chapter) in 1613, when results were presented in books rather than scientific journals. How did this fellow and his immediate successors, who included Isaac Newton, cope? The key to this question is to find out how many scientists actually contribute to the "multiplicity of books," or in today's terms the rapid growth in the number of scientific papers.

Analyses of the contents of numerous scientific journals indicate that science is a very undemocratic activity. If we regard the number of scientific publications as a measure of productivity, then we find that a few scientists do a disproportionate amount of the work while large numbers do a bare minimum. This result can be expressed mathematically: the number of authors producing n papers is inversely proportional to the square of n. This statement has the following approximate translation. A small number of scientists, N say, do most of the work; they write approximately half of all the papers that are published. The rest of the papers are produced by more than $N \times N$ scientists. Thus, if the small, very active group totals ten, then the much larger, lethargic group has more than a hundred members. This empirical law, which is sometimes referred to as Pareto's law, happens to fit a variety of phenomena. For example, if there is an increase in the national wealth, then a few very rich people get a disproportionately large share of that increase, while large numbers of relatively poor people get a very modest increase. Similarly, if the population of the United States doubles, then not all towns and cities double in size. Instead, a small number of large cities grow enormously, while a large number of small cities grow slightly.

In the same way that we congregate in large cities, rather than increase the population density uniformly across the country as our numbers increase, so scientists, as their numbers rise, tend to congregate mainly in a few disciplines, institutions, and even in the use of certain journals. (If there are a hundred scientific journals in a library, readers are likely to consult ten of those far more often than the others; out of a thousand journals, approximately thirty will prove most popular.) This result has interesting implications. If an institute wishes to acquire five highly productive scientists, it will have to hire twenty-five scientists, unless the institute is highly discriminating in whom it appoints. If the institute or university wants to produce ten very productive new scientists, it has to train one hundred students.

The number of scientific papers being published is a crude measure of productivity; it gives equal weight to one paper of Einstein on relativity and to a trivial paper by an unknown author on an obscure topic that interests no one. Furthermore, this measure has the unfortunate consequence of encouraging people to publish simply for the sake of having a long list of publications. A preferable measure of scientific productivity takes into account how often a paper is cited in other publications. (The references made to various publications are usually listed at the end of a paper.) Analyses of these references in large numbers of papers in various journals once again confirm that science is undemocratic; once again a Pareto law is appropriate. A great many published papers are cited once or not at all; a small number of papers receive enormous attention. In general, if the important papers are N in number, then the total number of papers is on the order of $N \times N$. Thus, of the first ten papers in a new field, three will be cited very often; of the first one hundred papers, ten will have that distinction; of the first one thousand, approximately thirty will prove particularly valuable. Making important scientific discoveries is apparently similar to picking apples in an orchard.[3] The first people to arrive in the orchard have little trouble finding choice apples. Those who arrive late generally have poor pickings, unless they are very talented at spotting what most others fail to see. This result explains why the people who first started studying El Niño are cited often; Walker and Bjerknes wrote important papers about the Southern Oscillation and El Niño not only because they were brilliant scientists but also because they were among the first to investigate those phenomena.

Today, we know so much about El Niño that very few of the huge number of papers devoted to him add significantly new information. Paradoxically, this state of affairs makes it relatively easy to obtain funds to study El Niño and to publish papers that elaborate slightly on some well-known feature of his. The scientific community tends to be conservative; it is easier to obtain funds for pedestrian investigations of phenomena that are widely accepted as being important than for explorations of exciting but tentative, speculative ideas concerning phenomena that are poorly understood. Young scientists quickly learn that the best strategy for publishing as many papers as soon as possible is to embroider on an established and accepted result rather than to introduce an original and controversial idea.

The number of highly productive, influential scientists in a field grows at a rate that is slow in comparison with the explosive growth of the total number of scientists in that field. Hence the communication crisis associated with rapid growth can effectively be solved, and the advancement of a field can be facilitated, by having the relatively small, influential group of scientists form an "invisible college" that promotes close interactions between its members. (Those members are generally affiliated with different institutions.) One of the first groups to do this, in Great Britain in the mid–seventeenth century, later organized itself into the Royal Society of London. Today it is still common, in highly competitive fields, for the scientific elite of a specialty to form an "invisible college." The members circulate preliminary reports on new results among one another and stay in close contact by becoming affluent commuters who meet frequently in select conferences and at prestigious institutions.

The relatively small number of people studying El Niño and related phenomena in the 1970s and 1980s formed such an invisible college. During that exciting period of rapid advances, those scientists promptly reported their results in a highly successful newsletter for their group, and they participated in lively discussions and debates at semiannual, informal meetings at a succession of different venues. The meetings proved so successful and popular that the number of people desiring to attend increased steadily. The number of scientists doing most of the significant work remained relatively small, so that maintenance of a relatively small invisible college was still feasible. However, certain developments in the early 1980s caused the study of El Niño to be transformed from "small" to "big" science, introducing new problems related to the management of science.

After the devastating El Niño of 1982, that phenomenon became important not only to scientists but to the world at large. This meant that scientists had to accept the responsibility of keeping the public informed, on a routine basis, of the whereabouts of El Niño. Furthermore, during the 1980s, not only El Niño but the global climate in general, and the possibility of global warming in particular, became matters of public concern. These developments called for a change in the way scientists organized their activities, because such serious affairs cannot be left in the hands of small, informally organized groups of select scientists who form invisible colleges and who are unfamiliar with the corridors of power. To bridge the worlds of science and of human affairs requires science managers to help scientists organize their activities and to secure resources for scientists by persuading government officials and funding agencies of the social importance of that branch of research—in this case, the importance of predicting El Niño and global warming. The successes of this new arrangement are numerous and include the following: a huge increase in public awareness of El Niño and global warming; new resources for scientific investigations of a wide range of phenomena related to weather and climate; funds for the deployment and maintenance of the TOGA array of instruments, which continually monitors conditions in the tropical Pacific Ocean; and the establishment of numerous national and international committees that produce such a plethora of newsletters, and organize meetings with such frequency, that there is no shortage of forums for the discussion of El Niño and related phenomena. At times these activities appear so frenzied—everybody serves on too many committees, attends too many meetings—that the science seems to be overorganized. Is frenzied activity being mistaken for accomplishment?

At present, a strong commitment to democratic principles governs the organization of scientific activities related to El Niño. In principle, this is a fair and just arrangement, except that science is not democratic. As mentioned before, the number of original and highly productive scientists in a field grows far more slowly than does the total number of scientists. In the words of de Solla Price, "The number of giants grows so much more slowly than the entire population that there must be more and more pygmies per giant, deploring their own lack of stature and wondering why it is that neither man nor nature pushes us toward egalitarian uniformity."[4] This statement is only partially correct because, once "small science" becomes "big science," the "pygmies" have ample opportunities to become influential. For example, if the reviews of a paper submitted for publication, or of a proposal submitted for funding, are treated democratically, so that each opinion carries equal weight, then the final decision will reflect mainly the opinion of the "pygmies" because they are the majority. The first step to a compromise between the conflicting demands of "big science" and "small science" is to recognize that there are serious conflicts.

How much longer can the remarkably rapid growth in the number of publications devoted to El Niño continue? Such growth in a subdiscipline of a certain field is not uncommon in science. In the 1960s and 1970s the number of papers on plate tectonics, the theory associated with continental drift, exploded in a similar manner.[5] Today the growth in that field is significantly slower; other branches of the geosciences are now attracting more attention and are growing at a more rapid rate. Scientific interest in El Niño is likely to continue growing for the time being because, despite the enormous progress in our understanding of that phenomenon, some of its major features remain unexplained (see chapter 15). Of special interest is the possibility that the current rise in the atmospheric concentration of carbon dioxide will result in perennial (rather than intermittent) El Niño conditions. For an assessment of this possibility scientists need to investigate certain climate changes that occurred in the distant past. (Chapter 16 has a brief description of those changes.)

The use of the geological records of past climates as tests for climate models that anticipate future global warming, will require close collaboration between two communities that have had relatively little interactions thus far. One is the community of paleoclimatologists; the other is the group of scientists whose concern is the weather and climate of today. The latter group is now turning its attention to climate variability over such long periods, of decades and more, that instrumental records to test theories and models are unavailable. The only way to overcome this serious problem is to turn to the geological record and to appeal to paleoclimatologists for assistance. Collaboration is in the interest of the latter group too. The detailed reconstructions of very different climates at different times in the past are most impressive and amount to a significant scientific achievement, but the lack of explanations for major phenomena such as the recurrent Ice Ages of the past million years, and the warm Cretaceous period when dinosaurs were the dominant animals, points to a need for testable hypotheses that can guide further data collection. The next few decades could be a period of exciting progress, from descriptions of past climates to explanations for those climates, provided that the two communities collaborate. This is easier said than done, because the activities of each community already amount to "big" science. Each has numerous national and international meetings, many journals, and countless committees that organize a variety of projects, including projects to promote interactions between the two communities. Some people find it difficult to see that there is a problem with communication, because they do not appreciate that "big" science, though necessary, is not sufficient for progress on the questions that have to be answered. For the study of global climates to enter a new and exciting phase it is of paramount importance that we foster the activities of "invisible colleges" — small, self-organizing, informal groups of scientists who practice "small" science and who bring fresh insights.

In the foreseeable future, interest in El Niño will continue to grow in scientific circles, but in the popular press interest is likely to be cyclical. For the time being, El Niño appears every four or five years, a period sufficiently long for memories of the last visit to have faded by the time El Niño appears again, thus justifying renewed discussions of each and every aspect of this multifaceted phenomenon. Fortunately it is as inexhaustible a topic of conversation as is the weather. We will never be indifferent to El Niño.

Part 3 | Common Ground

SEVEN

The Perspective of a Painter

Painting is a science, and should be pursued as an enquiry into the laws of nature. Why, then, should not landscape painting be considered as a branch of natural philosophy of which pictures are but the experiments?

—John Constable (1776–1837)

An English teacher in Australia had enormous difficulty getting a group of aborigine children to draw in perspective. They insisted on drawing in profile, the way the ancient Egyptians did. The teacher was overjoyed when, after many attempts, one child finally produced a drawing in the desired style. The youngster, when asked to explain to his mates what they needed to do, told them: "She wants you to draw it the way it looks, not the way it is."

Culture strongly influences the way we perceive reality. That is why we can easily tell whether a landscape was painted by a Chinese or a Dutch master; that is why a cock crowing in the morning says "cock-a-doodle-doo" to the English, "cocorico" to the French, and "kikeriki" to the Germans.[1] In the case of El Niño, we at first regarded him as a blessing, now we consider him a disaster, and in future we could conceivably change our minds again and accept him as a godsend. Each of us perceives reality differently at different times, but we nonetheless accept that there is one reality behind shifting appearances. Scientists claim that they have access to that reality, that their shifts in opinion are not arbitrary. They insist that, with time, they advance toward a more accurate understanding of reality. Once

Figure 7.1. © The New Yorker Collection 1955. Alain from cartoonbank.com. All rights reserved.

they learned that El Niño has global dimensions, they abandoned earlier notions that it is a regional phenomenon confined to the shores of Peru and are very unlikely to return to those notions. Science progresses and provides us with increasingly realistic models of natural phenomena such as the weather and El Niño, models that explain and predict their behavior with greater and greater accuracy.

Alain, in the cartoon reproduced in figure 7.1, suggests that Egyptian artists produced pictures in their traditional style, in profile, even though they used the objective, scientific method of perspective drawing. Is Alain poking fun at so-called scientific objectivity? Is he questioning the claim of scientists that their methods produce objective results? A visit to a shop that sells television sets can help us answer this question. When the different sets in the shop are all tuned to the same television channel, so that they all show the same image, then exactly the same information (from cameras at a football field, say) reaches each television set. The engineers who made the different sets all have the same goal, namely, realism. The images on the screens from different manufacturers are nonetheless different: some screens seem to be painted in pleasant pastel colors; others give a more dramatic image because the contrasts between the colors are sharper. Which manufacturer is providing the true image?

The painter John Constable insisted that "painting is a science," that not only scientists but painters too explore reality by making observations, performing experiments, establishing the laws that govern natural phenomena, and using those laws to simulate reality. This statement merits close examination because, if correct, it enables us to discuss, in artistic terms familiar to everyone, scientific issues that many laymen find puzzling. For example, what do scientists mean by a "model" of El Niño, or of Earth's climate? Why do different computer models for the prediction of El Niño, or of future global climate changes, give different results even when each model is described as realistic by its creator? Parallels between the activities of painters and scientists can be of benefit to scientists too. They can help scientists address the questions raised by Alain's cartoon, questions about the degree to which subjective considerations influence scientific results. Scientists are reluctant to acknowledge that such considerations play a role, but disputes and disagreements among scientists often tell us otherwise.

Why Painting Can Be Regarded as a Science

Realism in pictures, although it was of secondary importance to artists in ancient Egypt and during the Middle Ages, became a primary goal during the Renaissance, when artists attempted realistic depictions of three-dimensional scenes on flat surfaces.[2] They quickly discovered that perspective drawing requires the identification of a distant vanishing point onto which receding lines converge. Because of those lines, we see the three figures in figure 7.2 as a small, an average, and a large person. Actually, the three

Figure 7.2. From Metzger[3]

figures on paper have exactly the same size. (Check with a ruler!) An artist who wishes to draw any object or collection of objects in perspective has to distort their shapes, to a degree that depends on their distance from the observer.

In figure 7.2, the convergent lines that provide a sense of depth are explicit. In the cartoon by Alain, those lines are introduced more subtly by covering the floor with square tiles whose edges are converging lines. The cartoon, in addition to black lines on white paper, uses shading to indicate shadows, behind the model and under the chairs for example. Such a use of chiaroscuro (literally light-dark) enhances the three-dimensional effect by implying that light shines onto the scene from a certain point, so surfaces shielded from the light are shaded. Gifted artists can create a three-dimensional effect by relying only on shading, dispensing with convergent lines altogether. A sense of depth can be enhanced further by means of sfumato (literally "turned to vapor"), a technique refined by Leonardo da Vinci (1452–1519). He observed that the outline of a close object is sharper than that of a remote one. Distant objects should therefore be blurred deliberately.

Color gives artists additional freedom to explore the representation of three dimensions on a canvas. For example, a careful gradation of hues in landscapes—from mellow browns in the foreground, at the bottom of the canvas, to cool silvery blues in the background, toward the top of the canvas—can suggest light and distance; a painter can create the illusion of three dimensions by using a certain range of tonal gradations. So successful was the painter Claude Lorrain in applying these techniques to his landscapes that it led to the invention of the "Claude glass," a curved mirror with a toned surface. When looking at a scene through that device, a painter saw the actual colors transposed into a narrower range of tones appropriate for a picture. To achieve the desired impression of light and depth on the canvas, the actual colors, the bright green of the grass in the foreground for example, often had to be toned down. Constable, who wished to be more truthful to the actual color of the grass, experimented with widening the range of tones. He achieved success and set the stage for later painters such as Corot to paint even brighter landscapes.

To give an illusion of reality on a canvas, an artist has to learn the complex rules that govern our interpretation of shapes and colors. The challenge is to conjure up a convincing image even though the relative sizes of objects are severely distorted and even though no individual shade on the canvas corresponds exactly to the color in reality. By a method of trial and error (by means of experiments) artists learn how the appearance of any color on a canvas depends critically on the adjacent colors and the relative areas they occupy. (Imagine a surface, uniformly red, with superimposed horizontal lines that are white near the top, black near the bottom. The top of the surface will appear to be a lighter shade of red than the bottom.)

By the end of the nineteenth century, artists had essentially solved the problem of representing three-dimensional reality on a flat canvas. Although the methods by which they learned the "laws of nature" have much in common with the methods of scientists, the arts and the sciences had nonetheless become sharply divided. That divide grew significantly in the wake of

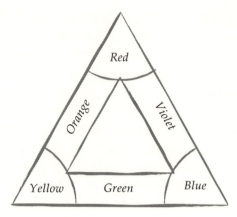

Figure 7.3.

Isaac Newton's pioneering studies of the phenomenon of color. Particularly troublesome to painters was the proposal that each color is distinct. (Each of the different colors in a beam of white light has a distinctive wavelength, the distance between consecutive wave crests. That distance is larger for the color red than for the color violet.) This notion proved problematic to painters, who knew for a fact that not all colors are equal, that a few are primary and hence of greater importance than the others: every other color can be created by mixing, in the right proportion, a few primary colors.[4]

For a while, the number and the identities of the primary colors were matters of dispute, but in due course it was established that there are three primary colors: red, yellow, and blue.[5] The painter Delacroix arranged them in a triangle (figure 7.3), from which it is obvious that yellow plus blue produces green, and so on.

Artists insist that there are only three primary colors, but scientists regard every color as primary. These different views were reconciled in the nineteenth century with the recognition that the retina of the human eye has only three types of light-sensitive cells, known as cones, for discerning color in bright light. (A fourth type of cell, known as a rod because of its cylindrical shape, enables us to see in dim light.) Each cone cell responds to one of three narrow bands of color centered respectively on the blue, green, and red wavelengths. Our perception of different colors depends on the combination of cone cells being stimulated. Because of our limited ability to discern different colors — animals with more than three types of cone cells do better — it is relatively easy to make a color television set for humans. The screen displays the colors of only three kinds of phosphorescent materials that glow, each kind with its own distinctive color, when struck by an electron beam. Our eyes perceive the thousands of dots of three different colors in a small area as a uniform color.

Painting can be regarded as a science whose goal it is to explore how certain very special instruments, our eyes and brains, perceive the physical world. In pursuing that goal, artists unwittingly made significant contributions to neurology, long before that science came into existence.[6] They continued to do so in the twentieth century, when many artists abandoned the quest for realism in pictures. To appreciate the significance of their achievement we need to recognize that the brain does not passively receive information about the external world — it does not merely register the arrival of pictures — but actively participates in generating the visual image, according to its own rules and programs. Different parts of the brain perform different specialized subtasks related to color, form, motion, face recognition, etcetera, and that information is then composed into a unified visual image. In the same way that we have to learn how to speak a language at an early age, so, early in life, we have to learn how to see. Children who are born blind and who, as adults, acquire sight (because of the removal of cataracts for example) are incapable of making sense of the information that their eyes receive and have difficulty recognizing three dimensions, distance, and size. People who suffer damage to certain parts of the brain can lose the ability to see color, or three dimensions, or motion. Artists appear to be intuitively aware of these scientific results concerning the functioning of the brain. The paintings of the Cubists of the early twentieth century, Picasso and Braque for example, can be read as efforts to make explicit our implicit awareness of the two-dimensional images that each of our eyes receives simultaneously, which our brains convert into three-dimensional scenes. To focus attention on this aspect of our vision, the Cubist painters used a narrow range of colors. Not so the Fauve painters such as Matisse and Derain. They had different goals that included the use of colors to express emotion. The color that the brain assigns to a certain observed area depends on the colors of adjacent areas, as explained earlier. The different areas are defined by their shapes or form, so color and form seem to be inseparable. The Fauve painters attempted a separation, a liberation of color from form, by assigning to certain forms colors not usually associated with those forms. In doing so they anticipated the findings of neurologists that the brain deals with color and form separately.

Apparent Differences between Artists and Scientists

Artists, because they make observations and use a method of trial and error, achieve their goals in a manner that, in many respects, can be described as scientific. We obscure the considerable parallels between the arts and the sciences by focusing mainly on the manner in which results are presented, in the case of the arts, and on the nature of the results rather than their presentation, in the case of the sciences. Scientists emphasize that their results are universal — independent of race or culture — but an artist's creation is usually seen as intensely personal. We confidently state that if Leonardo da Vinci had not existed, the Mona Lisa would never have been painted. If, on the other hand, Einstein had not existed, some other scientist would have thought of the theory of relativity. That theory provides no evidence that Einstein was German, nor does Newton's law of gravity indicate that he was English. However, different artists produce very different pictures of the same scene; each picture tells us volumes about the culture and personality of the artist.

These arguments in response to Constable's claim that painting is a science are misleading because they fail to distinguish between an idea and the presentation of that idea, which can take the form of a drawing on a sheet of paper or an article published in a scientific journal. Establishing a law of nature is one thing; a separate matter is the use of that law to explore reality, either by providing a scientific explanation for a phenomenon or by producing a picture. We make this distinction between an idea and its presentation when we refer to a certain artist as "a painter's painter." In the work of such an artist the principles invoked to depict a scene are of more interest than the scene itself. Thus the work of an artist who first discovers how to draw in perspective can be of interest mainly because of the novel methods rather than the specifics of the pictures produced. There is no need for the term a scientist's scientist, because scientists very seldom write for the public; they write almost exclusively for other scientists. Many are uninterested in, or are incapable of, explaining their scientific results to the public; that service is provided by science writers, some of whom have very little scientific training. (Newspaper editors insist that their art or music critics have a background in art or music, but science, unfortunately, is often treated differently.)

In a paper written by a scientist, all that presumably matters is the idea being expressed, not the manner in which it is presented. When presenting a paper at a seminar or meeting, a humanist usually reads from a prepared text — the words are carefully chosen and are of the utmost importance — but a scientist improvises, often inarticulately. To scientists, the words seem to be of minor importance — scientists tend to show a great many figures and graphs. Only the message matters. The physicist Ludwig Boltzmann nonetheless observed:

> Even as a musician can recognize his Mozart, Beethoven, or Schubert after hearing the first few bars, so can a mathematician recognize his Cauchy, Gauss, Jacobi, Helmholtz, or Kirchhoff after the first few pages. The French writers reveal themselves by their extreme formal elegance, while the English, especially Maxwell, by their dramatic sense. Who, for example, is not familiar with Maxwell's memoirs on his dynamical theory of gases? . . . The variations of the velocities are, at first, developed majestically; then from one side enter the equations of state; and from the other side, the equations of motion in a central field. Ever higher soars the chaos

of formulae. Suddenly, we hear as from kettle drums, the four beats, "put $n = 5$." The evil spirit V (the relative velocity of the two molecules) vanishes; and even as in music, a hitherto dominating figure in the bass is suddenly silenced, that which had seemed insuperable has been overcome as if by stroke of magic. . . . This is not the time to ask why this or that substitution. If you are not swept along with the development, lay aside the paper. Maxwell does not write program music with explanatory notes. . . . One result after another follows in quick succession till at last, as the unexpected climax, we arrive at the conditions for thermal equilibrium together with the expressions for the transport coefficients. The curtain then falls!

The cultural background and even the identity of an author can be inferred from the manner in which results are presented, in the entirely objective and highly abstract field of mathematics as much as in art. Chinese and Dutch masters produce pictures in styles that reflect the subjective canons of beauty of their culture. For example, western Europeans usually make a clear distinction between the observer and the observed, an attitude that readily finds expression in perspective drawing: the scene portrayed within a frame is separate from the viewer of that scene. Oriental culture encourages the artist to be at one with the scene being depicted, so clues as to the artist's location in the picture are minimized. Shadows, for example, are discouraged because they enhance the illusion of perspective. In different countries, youths are trained differently, in ways that reflect the different ways that cultures perceive beauty. This is true of the training of artists and also of scientists. It is evident in the way the great Isaac Newton interpreted the results from his experiments with prisms.

In 1666, at the age of twenty-three, Newton demonstrated that white light is composed of the sum of all the colors in the rainbow. He described his experiments as follows: "In a very dark Chamber, at a round Hole, about one third part of an inch broad, made in the Shut of a window, I placed a Glass Prism, whereby the beam of the Sun's Light, which came in at that Hole, might be refracted upwards toward the opposite Wall of the Chamber, and there formed a colored image of the Sun." In his writings and illustrations, Newton arranged the colors of the spectrum of sunlight in a circle, with violet next to red. The circle appealed to painters because it made visual sense; our eyes and brains perceive the colors violet and red as being close to each other. To physicists, however, there is no justification for a circle with red next to violet, given that red refracts more than violet does. (In terms of wavelengths, violet waves are far shorter than red ones.) Newton nonetheless drew a circle of colors because he was searching for parallels between light and sound. When he first passed white light through a prism, he was able to count eleven different colors. At a later time he decided on five before he finally added two more, orange and indigo, to arrive at seven colors. These various choices for the number of colors in the spectrum are all arbitrary; in reality there is a very huge number of colors between red and violet (that is, between the longest and shortest visible wavelengths of light). Newton's final choice of seven was an attempt to establish an analogy between visual and aural phenomena. In the diatonic scale, an octave is divided into seven musical tones; the eighth is the same as the first, at a higher pitch. Isaac Newton, the prince of objective scientists, allowed a certain sense of beauty — a desire for parallels between sight and sound — to influence his interpretation of visual information from a prism!

Style provides clues about the culture of a scientist or artist. So does substance. Whether a painter chooses as subject a madonna and child, a haystack, or a can of soup depends on the period and country in which the artist lives. Science too is influenced by historical factors. Consider the study of atmospheric storms. During the industrial revolution, a desire to improve the efficiency of steam engines motivated scientists to explore the concept of energy. Meteorologists then applied those ideas to the atmosphere. When they learned that the heat energy required to convert water into steam (a gas) is not lost but can be recovered by converting the gas water vapor back into liquid water, they proposed that clouds be regarded as heat engines: the conversion of water vapor into liquid water in a cloud releases energy that drives winds.

Such ideas are relevant to some storms, especially hurricanes, but fail to explain most storms in midlatitudes. To explain the latter phenomena, Scandinavian meteorologists (led by Vilhelm and Jacob Bjerknes, father and son), early in the twentieth century, introduced new ideas and terminology in keeping with the contemporary news of World War I battles farther south in Europe. These meteorologists proposed that the atmospheric storms in high latitudes develop when masses of air from different parts of the globe go on the march and collide along warm or cold fronts. The collision of warm, moist air from maritime regions in low latitudes with cold, dry air from continental regions in high latitudes can result in the release of a considerable amount of energy that manifests itself as storms.[7]

The verification of scientific ideas, concerning the generation of storms for example, requires two steps: the formulation of theories, sometimes expressed as mathematical equations, and the acquisition of measurements to check the theories. Some people are so engrossed in the mathematics that their studies of El Niño, say, seem to be mainly exercises in deriving and solving certain equations. To others the challenge is statistical, providing an accurate description of the sea surface temperature patterns associated with El Niño and La Niña on the basis of sparse measurements with random errors. Every investigator looks at a complex phenomenon such as El Niño from a different perspective, and in their studies each displays a particular penchant for statistics, or mathematics, or insightful, qualitative arguments involving physical principles. In the same way that different artists see the same scene differently, so different scientists have different perspectives on the same complex phenomenon. All have a tendency to believe that their own perspective (or the tool or model each one developed) is the one that matters most.

The degree to which subjective considerations influence the work of scientists or artists is minimal when they are dealing with simple problems. A perfect white cube on a featureless flat floor, an isolated pyramid on level desert sands — these leave little room for cultural biases. Different artists will produce essentially the same picture, irrespective of their background. In the scientific arena, the scope for subjectivity is similarly limited when the phenomena being explored are relatively simple and lend themselves to replicable experiments in a laboratory. A complex phenomenon, however, is another story because the artist or scientist is then obliged to make numerous choices, on the basis of a personal philosophy or a subjective concept of beauty. As mentioned earlier, a visit to a shop that sells television sets reveals the choices engineers have to make when trying to portray reality.

Constructing a Model of Earth's Climate

In the same way that artists use rules (those for perspective drawing for instance) to create a picture of a scene, so scientists use certain laws to simulate natural phenomena. If the phenomenon is simple—the orbit of a single planet around a sun—then application of the rules (the laws of nature) is straightforward. In the case of a complex phenomenon, however, the scientist proceeds by first considering highly idealized versions (of the weather, say), then gradually adding elements that initially were neglected, thus developing a hierarchy of models, from simple to elaborate. The goal is to gain an understanding of the phenomenon before attempting a realistic simulation. This procedure has a counterpart in the world of artists. An artist who embarks on a project that culminates in a realistic oil painting of a landscape is likely to start with a quick pencil drawing that captures the essence of the scene, then a few watercolors of certain key elements. In Constable's case, the preliminary studies include a full-size, spontaneous sketch that shows "amazing bravura of handling the heavily loaded brush-strokes and pigments dashed on and worked with a palette-knife."[8] Constable's final painting is a contemplative, tranquil recollection of the landscape. In this hierarchy, the set of rules gets progressively more complicated. A pencil drawing requires familiarity with a relatively small number of rules. The production of a watercolor involves the rules concerning the effect of one color on an adjacent one. As the picture becomes more realistic and detailed, the artist invokes more complex rules.

To illustrate how scientists gain an understanding of a complex phenomenon by developing a hierarchy of models, let us consider the climate of our planet. We can mimic the scientific procedure by pretending that we are from another galaxy and wish to build a habitable planet. We start with a bare-bones model (to be known as Model I) and step by step add "options" that increase its habitability. The increasingly sophisticated models are referred to as Models II, III, IV, and so on.

Model I: A Barren Rock

We, the visitors from another galaxy, observe that Venus, Earth's neighboring planet closer to the Sun, is too hot to be habitable, while Mars, the neighbor farther from the Sun, is too cold. Earth is apparently at just the right distance from the Sun. Because temperature is of the utmost importance to us, we choose as our initial model for a habitable planet a barren spherical rock, the size of Earth, with an orbit around the Sun that coincides with the orbit of Earth. To our dismay we quickly discover that a barren rock at the "right" distance from the Sun turns out to be much too cold for our comfort. Its globally averaged temperature is $-18°$ C, whereas the corresponding temperature of Earth is $+15°$ C. We consult a scientist to find out where we went wrong.

To solve our puzzle, the scientist invokes the laws of nature. In general, these laws identify invariants and thus express the thoughts of the Greek philosopher Democritus, of around 400 B.C., who stated that "nothing can arise out of nothing; nothing can be reduced to nothing." For the problem at hand, determining the temperature of a planet, the relevant invariant is energy; it can neither be created nor destroyed. In the same way that the rules for perspective drawing enable an artist to depict a huge variety of scenes, so the seemingly simple principle that energy is conserved can yield an astonishing amount of information about natural phenomena, such as the climates of different planets.

The sunlight shining on a planet heats it up. As its temperature rises, the planet radiates more and more heat into space until, in a state of equilibrium, the heat gained equals the heat lost. (The latter statement expresses the law for the conservation of energy.) From measurements of the sunlight incident on a planet, and from information about the amount of energy that bodies at various temperatures radiate, the theoretical temperatures of the different planets can be calculated:

Mars	$-53°$ C	$-56°$ C
Earth	$15°$ C	$-18°$ C
Venus	$430°$ C	$41°$ C

The temperatures in the first column are those actually measured at the surface of each planet; those in the second column are the theoretical temperatures that calculations yield for a barren rock at the same distance from the Sun as that planet. The calculations confirm our earlier findings for a barren rock in Earth's orbit, and they reveal discrepancies for the other two planets too.

The scientist explains that the critical difference between the barren rocks and the three planets we have been considering is each planet's atmosphere. The transparent, gaseous blanket that envelops each planet provides a greenhouse effect. The discrepancy between the calculated and observed temperatures — between the two columns in the table — is a measure of the greenhouse effect. It is minimal on Mars and enormous on Venus (whose atmosphere is composed almost entirely of the greenhouse gas carbon dioxide). On Earth we are the beneficiaries of a greenhouse effect that amounts to a warming of $33°$ C! To make our barren rock habitable, we need to provide it with an atmosphere.

Model II: A Rock with a Static Atmosphere

We order an atmosphere for our rock and carefully specify that its composition — its concentration of different gases — should be the same as that of the atmosphere of planet Earth. We now expect the temperature at the surface of our model planet to be comfortably warm. The scientist predicted the temperature of the barren rock very accurately, so, before proceeding, we apply his methods to determine the temperature of a rock with an enveloping atmosphere. Our calculations are again based on the law for the conservation of energy, but this time we apply the law repeatedly to a succession of spherical shells into which we divide the atmosphere. This exercise yields a temperature for each shell, thus providing us with a temperature profile for the atmosphere, a description of how temperatures vary with height above the planet's surface. Apparently temperatures will be low at great altitudes, as low as on the barren rock, but at the surface they will be far warmer. The atmosphere will indeed provide a greenhouse effect. However, the calculations indicate that it will be too much of a good thing. Globally averaged temperatures will be 67° C at the surface. When we alert the scientist to this potential problem, he nonetheless advises us to go ahead and to acquire an atmosphere similar to that of Earth. He promises us that temperatures at the surface will be lower than our calculations indicate.

Model III: A Rock with an Atmosphere in Motion

The moment our spherical rock acquires the specified atmosphere, we find that the very hot air at the surface starts rising spontaneously. This phenomenon, known as convection, redistributes heat vertically, cooling off the surface while warming up the higher elevations. In addition to this vertical redistribution, there is also a horizontal redistribution of heat: winds start to blow, carrying warm air from the very hot tropics to the much cooler polar regions. Our calculations were incorrect because they were based strictly on the law for the conservation of heat energy and failed to take into account the winds, which are governed by additional laws, those for the conservation of momentum, for example. Model III therefore requires more elaborate calculations. One way to proceed is to divide the atmosphere into spherical shells as before and also into latitudinal bands. By taking into account that the equatorial band receives more heat than the neighboring bands farther poleward, it is then possible to cope with motion from one band to the next.

Model IV: A Planet with an Atmosphere and Ocean

Our planet with an atmosphere in motion is not as hot as the one with a static atmosphere, but temperatures are still too high. The scientist recommends an ocean for our planet. He explains that water has very remarkable properties and will be of benefit to us in innumerable ways. Among others, it will cool the surface of the planet when the winds cause evaporation, transforming liquid water into the invisible gas water vapor. In the upper atmosphere, the condensation of that vapor into the water droplets of clouds is accompanied by a release of heat, heat originally taken from the ocean during evaporation. Hence the presence of water on a planet provides an additional means for redistributing heat vertically in the atmosphere. A mathematical model that copes with the hydrological cycle, by incorporating the law for the conservation of moisture, is capable of anticipating how the winds distribute moisture across the globe, and it can thus simulate a variety of climatic zones. To do so the model has to take a new complication into account: the presence of continents, whose thermal properties are different from those of the oceans.

We are now dealing with Model V. Further refinements could include an ocean that is in motion and that interacts with the atmosphere, thus allowing El Niño and La Niña to appear. The more sophisticated the model, the more demanding are the calculations. The development of some of these mathematical models therefore had to wait until electronic computers became available in the 1950s. As computers grow in size and power over the next several decades, so will the realism of the models known as General Circulation Models.

In this hierarchy, each successive model is more realistic than the previous one and involves the governing laws of nature in a more elaborate manner, thus requiring more and more calculations. Although the most elaborate models are the most realistic, the simpler ones are invaluable for providing an understanding of the phenomenon being investigated. If a model were so realistic that the data it provided were indistinguishable from actual measurements, then we would have a powerful tool for the prediction of weather, but we would still lack explanations for its underlying phenomena. For example, the main result from Model I, namely, that the barren rock is too cold, introduces the concept of a greenhouse effect and provides us with a quantitative estimate of that effect. A comparison of the results from Models II and III tells us about the importance of atmospheric motion to temperatures on Earth. Only from the sum of all the models do we gain an understanding of Earth's climate.

Although the most complex model is the most realistic, it is not necessarily the one that is most admired. This is true of models of weather and climate and also of paintings. Today many connoisseurs admire Constable's preliminary sketches more than his final paintings. Such subjective preferences have been debated since antiquity. Can line drawings be adequate representations, with color an inessential adjunct? The Romans regarded design as superior to color, which they associated with luxury. During the Italian Renaissance, artists from Venice tended to give supremacy to coloring, those from Florence to drawing. In the nineteenth-century version of this debate the artists Ingres and Delacroix represented the opposing poles even though both were supreme colorists. Sometimes, in some regions, one point of view prevails. In France at the time of Louis XIV, the Neoplatonic belief that the ideal beauty could be captured only in drawing—essential shapes uncompromised by matter—held sway. According to this belief, the ultimate drawing is on the surface of the earth, by means of garden plants cultivated to display, not the ephemeral appearance of nature, but its underlying order embodied in ideal geometric shapes such as the circle and square. The French classic gardens, at Versailles for example, are beautiful expressions of these ideas.[9] We can only speculate how Louis XIV would have responded to a pretty English cottage garden where colorful foxgloves, lilies, cosmos, and dahlias grow in profusion.

In the sciences, debates similar to those in the arts concern the relative merits of simple and complex models. Some people aver that far more challenging and rewarding than the development of a realistic model requiring the brute power of a computer is the formulation of a simple, idealized model that captures the essence of a complex phenomenon and requires the use only of pencil and paper. Those who attempt realistic simulations by means of complex, supercomputer models protest, insisting that theirs is the final word, the only one that matters.[10] The flaw in such a claim becomes evident when we consider how we would proceed should there be a model capable of perfectly realistic simulations. The results from such a model, as in the case of measurements made in reality, would be impossible to describe or interpret without the insights, concepts, and vocabulary gained from simpler, idealized models.

A simple model can capture the essence of the phenomenon under consideration by filtering out unnecessary details. Because of the high degree of idealization in such a model, its results have to be checked for consistency. Are the neglected factors indeed negligible? The error of providing apparent explanations for certain phenomena by means of highly idealized models, without checking whether the neglected processes are indeed negligible, is common — in the debate about global warming, for example. Critics of complex climate models often identify a process that happens to be absent from those models — typically a process involving some aspect of clouds. They then study the effect of that process in isolation from all the others that influence climate. If the process has a tendency to cool the planet, say, the critics call into question the results from the complex model that predict warming. This verdict is premature. One cannot be sure whether the isolated process is indeed important until the complex models incorporate the neglected process and determine its effect in the context of the great many others that influence climate.

In this respect the arts and sciences are very different. In the arts, an engraving, or a French classic garden, can be an end in itself, not necessarily the prelude to a grander work. Science, on the other hand, is a "perpetual becoming." Once scientists have identified the laws of nature, they next explore the consequences of those laws and continually expand the domain of validity of those concepts. A scientific result is always tentative, subject to revision should new information become available. Nobody who has set eyes on Vermeer's *View of Delft*, or some other great work of art, would describe it as a tentative statement that may have to be revised.

"Parameterizations" in Paintings and Computer Models

Attempts to depict certain phenomena on a canvas, or to simulate them by means of computers, sooner or later bring artists or scientists to an impasse. How, for example, can a painting give the impression of motion? Diego Velázquez was one of the first artists to give a realistic impression of a spinning wheel by introducing a blurring effect that simulates something whizzing across the field of vision. A few centuries later, when Duchamp painted *A Nude Descending a Staircase*, he suggested motion by means of some twenty abstract pictures of the nude in successive acts of descending. There clearly is no unique solution to the problem of depicting motion on a single canvas. Scientists have to deal with similar problems. Consider the winds that drive motion — waves and currents — in the ocean. The winds affect the ocean by generating waves at its surface, the whitecaps seen on a windy day. Each lasts a few seconds and has dimensions that can be measured in meters or less. In the course of a year, an extremely large number of them appear in the vast Pacific Ocean, which is 15,000 kilometers wide. How can each and every one of those waves be taken into account if the phenomenon of interest, El Niño, involves the entire Pacific and lasts for a year and longer? It will be a very long time before even the largest computer can cope. Scientists therefore have to create an "illusion" of the waves in order to simulate the effect of wind on the ocean. This can be done by including the waves implicitly rather than explicitly. The wind uses the waves as an intermediary to drive the oceans, so that the role of the waves is implicit when the wind above the ocean surface is treated as if it were a force exerted, in the direction of the wind, in the uppermost layer of the ocean. Such procedures for coping with phenomena that models cannot explicitly include are known as "parameterizations." A test for a parameterization, of surface waves in this example, is to have a model simulate the oceanic aspects of El Niño. Can the model reproduce how the complex system of eastward and westward currents and undercurrents in the upper layers of the tropical Pacific Ocean changed during a specific El Niño, the one of 1982–1983, say? Given the changes in the surface winds, the models do indeed prove capable of reproducing realistically the observed oceanic changes, which attests to the adequacy of the parameterization of the turbulent processes that translate the surface winds into forces that drive the ocean. On the basis of such results we can be confident that the models are accurate for a certain region — the upper layers of the tropical Pacific — over a period of several years. We will have to devise different tests to determine whether the models are reliable for the world ocean, from the surface to the ocean floor, over periods of several centuries.

The biggest challenge in the development of atmospheric models is coping with clouds. Artists too find clouds, among the most ephemeral and elusive phenomena of the visible world, a major challenge. The difficulty of judging size and distance in the sky, in the absence of a familiar reference such as a tree, makes application of the rules of perspective almost impossible. Furthermore, any attempt to depict a cloud makes it explicit that the result of an artist's labors depends not only on the artist but on the viewer too. Shakespeare captured the fantasies that clouds permit in the following exchange:

HAMLET:
Do you see yonder cloud that's almost in the shape of a camel?
POLONIUS:
By th' mass and 'tis, like a camel indeed.
HAMLET:
Methinks it is like a weasel.
POLONIUS:
It is backed like a weasel.
HAMLET:
Or like a whale.

POLONIUS:
Very like a whale.
                    William Shakespeare, *Hamlet*

Not only Hamlet, but some scientists too, make mischief by means of clouds. On some occasions those scientists argue that clouds cool the earth because they reflect sunlight; at other times they assert that clouds warm the earth because the water vapor in the clouds is a powerful greenhouse gas. Clouds do indeed play this dual role and are the biggest contributors to uncertainties in scientific results concerning future global warming. (Chapter 11 explores why these whimsical phenomena cause computer modelers so many woes.)

Mathematical models that attempt to simulate and predict weather and climate, the very elements of nature that have shaped our civilizations and that we hold in awe, are bound to arouse passions. Acquiring the considerable skills needed to cope with the sets of equations that correspond to a particular model of Earth's climate amounts to a significant investment in time and in emotions. The development of a realistic model that requires a powerful supercomputer is usually the coordinated effort of a team of scientists. This mode of operation is reminiscent of the workshops of the grand masters, where numerous assistant painters and apprentices each contributed to one or another aspect of a huge canvas. Scientists who deal with highly idealized models meet their challenge in an entirely different way. They usually work individually, do not require supercomputers, and solve their equations by analytical methods, using pencil and paper. Both this group and the ones developing realistic computer models invest so much time and energy acquiring their respective skills that they sometimes fall in love with their creations — models, instruments, techniques — the way the sculptor Pygmalion fell in love with his statue of a woman. Although the gods granted Pygmalion's wish and brought the cold stone to life, most artists accept that an abstraction can never be the true thing. For example, when someone said of a painting by Matisse, "But surely, the arm of this woman is much too long," the painter responded, "Ma-

dame, you are mistaken. This is not a woman, this is a picture."
Not all scientists are as objective. Some are unwilling to recognize
the limitations of their creations and insist that an idea or model
that clarifies a narrow aspect of reality solves the entire problem.
This can lead to heated debates such as the current one among
oceanographers concerning the role of different components of
the ocean circulation in climate changes. One group attributes
practically all climate changes to an altered thermohaline circulation (or oceanic conveyor belt), another questions whether there
is such a circulation, and a third emphasizes the wind-driven circulation of the upper ocean. Members of any of these parties
could be the student-artists in Alain's witty cartoon. The scientists studying El Niño are also engaged in a heated debate. Some
claim that that phenomenon is stochastic, others that it is cyclic.
To the first group the appearance of each El Niño is the response
to a definite "trigger," such as a random burst of westerly winds
along the equator. If that were the case, then El Niño would be
very difficult to predict. Others argue that El Niño is one phase of
a continual oscillation and can be anticipated far in advance. A
growing body of evidence suggests that El Niño, in reality, is a
cyclic phenomenon that is sporadically disturbed by random wind
fluctuations. The contentious scientists seem to have much in
common with the Australian children who maintain that their
drawings show things the way they are, not the way they look.

Scientists are making rapid progress with the development of
computer models capable of realistic simulations of Earth's climate. Of particular interest are detailed simulations of the climate changes that will accompany global warming over the next
several decades. At present the models produce reasonably accurate predictions for globally averaged surface temperature, but
the uncertainties in the predictions of specific climate changes
that different regions will experience are huge. How can those
large uncertainties be reduced? For guidance we can turn to the
meteorologists, who over the past few decades have developed
atmospheric models capable of consistently reliable weather predictions—for the next few days—for different parts of the globe.

The daily weather maps that appear in newspapers and on television are produced in a manner remarkably similar to the way an atelier produces large canvases depicting elaborate scenes. The different artists in a studio, after years of training, are all capable of realistic renditions of different objects — clouds, trees, the fall of drapery — and can thus contribute to the creation of new compositions that have such objects. A weather forecast similarly involves teams of scientists all schooled in certain fundamentals. They have used highly idealized models to study different weather phenomena in isolation — fronts, cyclones, hurricanes — and thus have an understanding of each of those phenomena. They have also used idealized models to create "road maps" that show the chaotic weather of our particular planet in the context of other possibilities, such as planets with no weather at all, or planets with weather that is very regular and hence highly predictable. This training provides meteorologists with a vocabulary to discuss, and with tools to analyze, a great variety of weather patterns, including those produced by the very sophisticated computer models of the atmosphere. Should a model produce an inaccurate forecast, the meteorologists are able to determine how it is flawed. Even more important, they know how to improve the model. Frequently the latter step requires improved "parameterization" of those processes and phenomena with which the model cannot cope explicitly. For example, in the same way that artists have to learn how to create the effect of leaves fluttering in the wind without depicting individual leaves realistically, so scientists have to devise techniques to reproduce the effects of innumerable clouds without simulating individual clouds realistically. A competent artist can depict winds that range from gentle breezes to stiff gales. A meteorologist has to represent in a model the very different clouds associated with a hurricane over the tropical Atlantic and with an onshore breeze that brings light rains to an island of Hawaii. To develop a "parameterization" of clouds sufficiently general to cope with such a broad range of possibilities requires careful observations of a great variety of possibilities. Opportunities for such observations come frequently because the weather changes often — and never repeats itself.

Climate changes can also cover a very broad range of possibilities. This is evident in the geological record of very different climates in Earth's past. In the same way that weather forecasting progressed as meteorologists learned to explain and simulate a diversity of weather patterns, so the reduction of uncertainties in the prediction of future climate changes will require explanations for, and an improved ability to simulate, past climates — the Ice Ages for example. The challenge to scientists is to avoid becoming like the students in Alain's cartoon, who interpret the world in terms of too narrow a range of possibilities.

EIGHT

The Perspective of a Poet

> There is no science without fancy and no art without facts.
> —Vladimir Nabokov (1899–1977)

Scientists, poets, and painters all search for order, for unity in nature's bewildering variety. Some people are under the false impression that scientists merely collect facts. They expect from climatologists, for example, a tedious recitation of when and how much it rains in different parts of the globe. Such facts are of course useful and can even be used to paint a vivid picture, as Ruskin demonstrates with a glorious torrent of words that describe the "variegated mosaic of the world's surface" (see the sidebar). To poets and scientists, however, facts are merely a means toward an end. Ruskin uses them in defense of gothic art. After the passage quoted in the sidebar, he asks us to contrast the "delicacy and brilliancy of color, and swiftness of motion," of creatures of lower latitudes "with the frost-cramped strength, and shaggy covering, and dusky plumage of the northern tribes." Ruskin states that, to appreciate the gothic artist, we should "not condemn, but rejoice . . . when, with rough strength and hurried stroke, he smites an uncouth animation out of rocks which he has torn from among the moss of moorland." To appreciate gothic art, we need to know the facts of geography; there is "no art without facts."[1]

A scientific description of climate in terms of strings of numbers that express how rainfall and temperature vary from place to place is dry as dust. Sir Gilbert Walker's strictly statistical descrip-

The Variegated Mosaic

We know that gentians grow on the Alps, and olives on the Apennines; but we do not enough conceive for ourselves that variegated mosaic of the world's surface which a bird sees in its migration, that difference between the district of the gentian and of the olive which the stork and the swallow see far off, as they lean upon the sirocco wind. Let us, for a moment, try to raise ourselves even above the level of their flight, and imagine the Mediterranean lying beneath us like an irregular lake, and all its ancient promontories sleeping in the sun: here and there an angry spot of thunder, a grey stain of storm, moving upon the burning field; and here and there a fixed wreath of white volcano smoke, surrounded by its circle of ashes; but for the most part a great peacefulness of light, Syria and Greece, Italy and Spain, laid like pieces of a golden pavement into the sea-blue, chased, as we stoop nearer to them, with bossy beaten work of mountain chains, and glowing softly with terraced gardens, and flowers heavy with frankincense, mixed among masses of laurel, and orange, and plumy palm, that abate with their grey-green shadows the burning of the marble rocks, and of the ledges of porphyry sloping under lucent sand. Then let us pass farther towards the north, until we see the orient colours change gradually into a vast belt of rainy green, where the pastures of Switzerland, and poplar valleys of France, and dark forests of the Danube and Carpathians stretch from the mouths of the Loire to those of the Volga, seen through clefts in grey swirls of rain-cloud and flaky veils of the mist of the brooks, spreading low along the pasture lands: and then, farther north still, to see the earth heave into mighty masses of leaden rock and healthy moor, bordering with a broad waste of gloomy purple that belt of field and wood, and, splintering into irregular and grisly islands amidst the northern seas, beaten by storm, and chilled by ice-drift, and tormented by furious pulses of contending tide, until the roots of the last forests fail from among the hill ravines, and the hunger of the north wind bites their peaks into barrenness; and, at last, the wall of ice, durable like iron, sets, deathlike, its white teeth against us out of the polar twilight.

— John Ruskin, *The Stones of Venice*

tion of the curious Southern Oscillation—the climate fluctuation involving vast masses of warm, moist air moving back and forth across the tropical Pacific—failed to sustain interest for long. Soon after he presented those results they fell into oblivion. For the statistics to acquire life, and for the beauty of the science to become evident, we had to wait until Professor Jacob Bjerknes proposed an explanation for the phenomena Walker documented. The science was incomplete until the facts, dutifully collected by the observer, found their complement in the fanciful hypotheses of the theorist. Bjerknes had several brilliant insights. First he recognized that the Southern Oscillation is the atmosphere's response to changes in the sea surface temperatures of the eastern tropical Pacific. Next he boldly stepped into the territory of oceanographers and argued that those oceanic temperature changes are a consequence of the wind fluctuations associated with the Southern Oscillation. Finally, in the most original part of his argument, he extricated himself from this circular reasoning by proposing that the ocean and atmosphere interact and spontaneously produce El Niño and La Niña. In the same way that we applaud Shakespeare for observing that

> the sails conceive
> And grow big-bellied with the wanton wind;
> William Shakespeare, *A Midsummer
> Night's Dream*

so we should recognize that the scientist Jacob Bjerknes was inspired when he proposed that the wanton wind flirting with the Pacific Ocean can generate children, El Niño and La Niña.

The Atmosphere

Let us, in the spirit of Ruskin, take a bird's-eye view of "the variegated mosaic of the world's surface," not in the Mediterranean, but instead in the eastern tropical Pacific Ocean. Today, commercial flights from Lima, the capital of Peru, to Panama City afford spectacular views of the different climatic zones in that part of the world.

Lima is perennially cloudy but has never had rain, at least not since Pizarro founded that city. It is located on a barren coastal plain between a cold ocean and the tall, snowcapped Andes. Immediately after takeoff from Lima, our airplane breaks through the shallow layer of unbroken clouds that stretch far westward, and we find the Andes rising steeply to the east. As we progress northward, the bare mountain slopes gradually gain some green, and by the time we reach the equator near Quito, the capital of Ecuador, the landscape is verdurous. Farther north the vegetation becomes even more lush and, in Panama, becomes a dense, tropical jungle. In the meanwhile, the clouds have changed. When the pilot starts the descent into Panama, he carefully chooses a path between towering cumulus stacks with ample blue sky between them. These clouds are in striking contrast to the flat, shallow, stratus deck near Lima. Passengers who continue the journey farther northwest along the coast, to Los Angeles, find that the climates that unfold are similar to those encountered on a return trip to Lima. It is as if the climatic zones are arranged symmetrically about the latitude of Panama City.

Panama City is at 10° N latitude. Why is the climate of the eastern Pacific symmetrical about that latitude rather than the equator? After all, sunlight is most intense at the equator and is perfectly symmetrical about that line. Another asymmetry reveals itself when we travel westward along the equator. The intensity of sunlight does not change with longitude, and yet we find ourselves moving from a region of shallow stratus clouds and low precipitation, along the shores of Ecuador and Peru, to lush tropical jungles in New Guinea and Indonesia, where cumulus clouds provide plentiful rainfall.

The key to these puzzles is the web of winds that distribute moisture across the surface of the earth. The westerlies and easterlies, the trades and the monsoons, are the invisible threads of the colorful climate tapestry of deserts and jungles, of steppes and tundra. Those winds are also the designers of the tapestry. They use a magical, colorless dye, water, to paint the landscape green (where it rains), white (where it snows), and "rich oriental colors" (where precipitation is withheld.) The winds harvest moisture over the oceans and store that dye in fantastically shaped granaries, clouds.

The design of the tapestry is strongly influenced by temperature patterns at the surface of the earth. Over the warmest regions in the tropics, hot air tends to rise into rain-bearing cumulus clouds, the surface winds converge onto those regions. That is why a sea breeze blows from the cold ocean toward the warm land. That is why the monsoons converge onto India in summer. Thus the puzzle of the climatic asymmetries in the tropical Pacific has a simple explanation: the asymmetries reflect surface temperature patterns. Cumulus clouds are prominent where sea surface temperatures are high, along the latitude of Panama (10° N approximately) and in the equatorial Pacific west of the date line. The air that rises there falls back to the surface over the cold waters of the eastern tropical Pacific, off Peru and California. In the latter regions, sinking air inhibits the formation of cumulus towers. Instead, the water vapor that evaporates from the cold ocean condenses into the low, horizontal layers of stratus clouds off California and Peru.

The idea that temperature patterns at the surface of the earth give rise to winds that design the climate tapestry can be tested: when the temperature patterns change, the tapestry too should change. That does indeed happen when El Niño occurs. The warming of the eastern equatorial Pacific is then associated with the appearance of rain-bearing clouds that convert the Peruvian desert into a garden. To understand changes in sea surface temperature patterns, we have to study the oceans.

The Oceans

Climatic zones as diverse as those on land characterize the oceans. Some regions have a dearth of marine plants and animals, others a plethora. The waters off the coast of Peru are exceptionally rich in marine life; anchovies and sardines are so plentiful that they attract, in huge numbers, larger fish such as yellowfin tuna, mackerel, sharks, and dolphins. The Great Barrier Reef off north-eastern Australia is an entirely different world. There, very warm waters wash over colorful coral reefs that provide a habitat for a multitude of exotic tropical species. To visit the marine counterpart of deserts on land, we have to voyage to the oceanic interiors, far from coasts and far from the equator. In satellite photographs of the chlorophyl concentration at the ocean surface, those vast regions are barren in comparison with the small, fertile coastal zones such as the ones off Peru and along the equator. Who designs the aquatic climate tapestry?

The factors that matter most to marine life are: (1) the availability of light, (2) the concentration of nutrients in the water, and (3) the temperature of the water. Sunlight, which readily penetrates through the atmosphere to the earth's surface, reaches only a few tens of meters into the ocean, so oceanic plants (which need light for photosynthesis) grow mainly near the surface. (They are known as phytoplankton, literally plants that wander.) Animals live off plants and off each other, so they too are found mainly in the upper ocean, in a shallow layer of warm, light-filled water that floats on a cold, dark, desolate abyss.[2] (The interface between the warm and cold waters, known as the thermocline, is particularly sharp and shallow in the tropics, where it is at a depth of 100 meters approximately. The average depth of the ocean is 4,000 meters.) The various life-forms consume nutrients (phosphates, nitrates, etc.), which therefore have a low concentration in the upper ocean. When the plants and animals die, they sink into the dark, cold ocean, where they decompose, so nutrients are plentiful at depth. Life consumes nutrients in the upper ocean, but death, followed by descent into the abyss and then decay, transports those chemicals to the deep ocean. To sustain life in the upper ocean, the cold water, rich in nutrients, has to rise back to the surface. This means that the ocean has to be in motion. Oceanic currents, the counterparts of atmospheric winds, create the marine climate tapestry.

Whereas the winds distribute moisture that falls from above, the currents distribute nutrients that rise from below. The winds respond to surface temperature patterns; the currents are driven by the winds. At first the surface waters flow in the direction that the wind is blowing, but then the rotation of the earth brings the Coriolis force into play, causing the water to drift to the right of the wind in the Northern Hemisphere, to the left in the Southern Hemisphere. What happens at the equator? To the north of that line, westward winds and the Coriolis cause the surface water to drift northward; to the south of that line, those forces cause the water to drift southward. To sustain this divergent motion, away from the equator, cold water, rich in nutrients, rises into the surface layers from below. As a result, the waters along the equator have an abundance of marine life. A similar parting of the waters occurs in certain coastal regions where the winds are parallel to the shore, and the Coriolis force causes offshore drift. These conditions are satisfied in "upwelling" zones off California and Peru and also off north- and southwestern Africa. All are rich fishing grounds. The California, Peru, Benguela, and Canary Currents then distribute the nutrients over wider areas.[3]

The oceanic climate tapestry depends on currents driven by the winds. A change in the winds changes the tapestry. During La Niña, when the trades are intense, the "upwelling" zones that are rich in nutrients are prominent. They practically disappear during El Niño, when the relaxation of the trades alters the currents in such a manner that they carry warm waters from the far west all the way to the shores of South America. This change in oceanic climate is devastating to the marine life in the eastern tropical Pacific, but fortunately the change is brief. Life has evolved to cope with such changes, the way it copes with the seasonal fluctuations.

The currents distribute nutrients and also create temperature patterns. The cold water that rises to the surface off Peru and California is swept westward. During the journey toward Asia and Australia, the water is heated by the sun. As a consequence temperatures are far lower in the eastern than western tropical Pacific. Along the western sides of the ocean basins, currents such as the Kuroshio and Gulf Stream carry warm water from low to high latitudes, giving the temperature patterns yet another dimension.

Ocean-Atmosphere Interactions

The atmospheric winds and oceanic currents that design the climate tapestries on land and at sea depend on each other. The winds drive currents that create temperature patterns that in turn influence the winds. The climate asymmetries of the tropical Pacific therefore involve interactions between the atmosphere, the oceans, and the continents. To develop an appreciation for those interactions, to gain an understanding for the factors that determine our climate, causing Panama to have plentiful rainfall, Peru very little, we have to follow the advice of Nabokov and become fanciful. In the same way that a poet has to inhabit different worlds—those of a cat or a horse in order to write about a cat or a horse—so a scientist has to imagine different worlds, those with and without ocean-atmosphere interactions for example. Let us first explore a world so idealized that it is perfectly symmetrical about the equator. In what respects will it still resemble our asymmetrical world?

Imagine a water-covered globe without any continents, heated by sunlight similar to ours, sunlight that is independent of longitude and is symmetrical about the equator. On such a planet, surface temperatures have a maximum at the equator, and the climate is symmetrical about that line. At the surface, moist air converges onto the equator and rises there. Aloft, the air, drained of its moisture, flows poleward until it subsides over the horse latitudes of the subtropics, perennially sunny regions with practically no precipitation. The surface winds, the trades, then return the air to the equator and replenish it with moisture by means of evaporation from the oceans.[4] This highly idealized world has a number of realistic features: rainy doldrums near the equator, where the southeast and northeast trades meet; marine deserts in the subtropics, where dry air subsides. This world, unlike ours, has maximum temperatures at the equator because temperature depends strictly on the local intensity of sunlight. Let us next move to a more complex world in which surface temperatures depend, not only on the flux of heat onto the ocean surface, but also on the winds and the currents.

In the water-covered globe considered thus far, the thermocline is so deep that the winds are unable to bring cold water to the surface. Let us next consider a similar world, except that the thermocline is shallow. In this new world, the winds should readily bring cold water to the surface at the equator because there the winds can drive currents that cause a parting of the surface waters. In principle this is possible, but not if the winds converge onto the doldrums along the equator and are very weak along that line. Let us briefly permit strong winds near the equator by introducing a modest perturbation that displaces the rainy doldrums to a region slightly off the equator, in the Northern Hemisphere, say. Now the winds from the south blow across the equator and penetrate into the Northern Hemisphere, as far as the displaced doldrums. The equatorial currents generated by these winds are such that cold water appears at the surface at, and south of, the equator. As a result, the displaced doldrums, which are always over the warmest water, remain north of the equator, where surface temperatures are high. This means that, in a world with a shallow thermocline, conditions that are symmetrical about the equator do not remain that way for long. Any temporary perturbation that causes a departure from symmetry has a permanent effect; the departure from symmetry persists! A symmetrical world with a shallow thermocline is unstable and must change.

The doldrums, if arbitrarily displaced away from the equator, remain in the new location. An arbitrary displacement is as likely to be northward as southward. Why, in reality, are the doldrums and the warm surface waters always north of the equator, near the latitude of Panama? To explain why Panama and not Peru has a warm ocean and rain-bearing clouds, we need to imagine yet another world, one with continents.

On a water-covered globe, the winds are westerlies in midlatitudes and easterly trades in the tropics. How are those winds modified by the presence of a continent that covers a band of longitudes and stretches from pole to pole, somewhat like the

Americas? Land has thermal properties different from that of water; sunlight can therefore cause a continent to get much hotter than the ocean. Along the equator, where the Coriolis force vanishes, the factors that give rise to a land-sea breeze cause the trade winds in low latitudes to intensify. Thus the presence of the Amazon Basin, and of the "maritime continent" of southeastern Asia and northern Australia, contributes to stronger trades over the Atlantic and Pacific.[5]

The presence of continents has another consequence. It affects oceanic conditions because the winds can pile up water against the sides of continents. Thus the trades over the Pacific pile up warm surface waters against the landmass in the west, causing the thermocline to slope downward to the west. Hence, because of continents, the thermocline in the tropical Pacific is deep in the west, shallow in the east. This means that the ocean-atmosphere interactions that convert a symmetric world into an asymmetric one, and that depend on a shallow thermocline, are confined to the eastern tropical Pacific (and Atlantic).

On a water-covered globe with a continent that is symmetrical about the equator, the asymmetry in the eastern side of the ocean basin could favor either hemisphere. The asymmetry could even flip-flop back and forth across the equator, so that the climate of Peru would sometimes resemble that of Panama and vice versa. This does not happen in reality, because the Americas are not symmetrical about the equator. Their western coast is inclined to lines of longitude so that San Francisco is farther west than Santiago in Chile. This arrangement of the continents, an accident of the drifting of the continents over millions of years, apparently contributes to temperatures that are higher off Panama than off Peru.[6]

The ocean and atmosphere, the two concentric, spherical shells of water and air that envelop our planet, have an intimate relationship. The ocean gains heat from the sun and transfers some of it to the atmosphere in the form of latent heat; this happens when water at the ocean surface evaporates. Moisture in the atmosphere is mostly in the form of an invisible gas, water vapor. That gas condenses into water droplets inside clouds, especially over the regions where surface temperatures are at a maximum in low latitudes. The heat that is released in the clouds when the vapor condenses into droplets is a major source of energy for the atmospheric circulation, for the winds. Those winds do not respond passively to sea surface temperature patterns but help create those patterns. They do so by bringing the ocean to life, by generating currents that circulate the waters, bringing cold water to the surface in some regions and dispersing that water over huge areas. The alluringly asymmetrical climate tapestry of our planet is the product of a complex interplay between the atmosphere, the oceans, and the continents. Thus far we have discussed how the interplay determines the static aspects of a tapestry. In reality the carpet is dynamic — its asymmetries intensify during La Niña, weaken during El Niño. We need the perspective of a musician to shed light on this dance to the music of time.

NINE

The Perspective of a Musician

Music creates order out of chaos; for rhythm imposes unanimity upon the divergent; melody imposes continuity upon the disjointed, and harmony imposes compatibility upon the incongruous.
—Yehudi Menuhin (1916–1999)

Upon opening her window one morning, an English lady on vacation in southern California was heard to complain about "another bloody sunny day." The lady was growing tired of predictable warm days followed by predictable cool evenings. To her, the daily fluctuations in atmospheric conditions in southern California are as monotonous as the music of a pendulum or tuning fork. The lady longed for additional notes—weather that changes continually so that each day is distinct. The weather in England is the music of an instrument capable of a tune. It is infinitely varied and has recurrent themes or leitmotivs that can be learned by watching the sky carefully: a ring around the moon is often a precursor of stormy weather; red skies at night are usually a shepherd's delight—the next day is likely to be splendid.

Scientists refer to the monotonous weather of southern California as "forced variability" of the atmosphere because it is related to daily and seasonal changes in sunlight in a direct and obvious manner. The far more subtle weather of England, where fluctuations in atmospheric conditions seem unrelated to changes in sunlight, is known as "natural variability" because it appears spontaneously, without any apparent cause. Its relation to music is more than metaphorical; the weather in places such as England can literally be viewed as the music of the atmosphere. In the same way that a musical instrument has modes of vibration that are easily excited—by plucking a taut violin string or blowing into a flute—so our atmosphere has natural modes of oscillation that are easily excited and that manifest themselves as weather phenomena.

Let us perform an experiment in which we excite the modes of oscillation of different objects by striking each of them gently, with a spoon. When we strike a table, a pot, a fine wineglass with a stem, a curtain, a bell—they all vibrate and produce sound, but the sound differs considerably from one object to the next. In some cases we hear cacophonous noise; in most cases the sound dies away quickly. The wineglass and the bell—and musical instruments in general—are exceptional because they readily produce beautifully harmonious music that lingers for a surprisingly long time. It is generally difficult to elicit a pure note from a solid object other than carefully designed musical instruments. Matters are very different when dealing with fluids, and therein lies the secret of our planet's music. Even a child can produce a pure note in a pond by merely dropping a pebble into the water. The perfectly concentric, expanding rings on the surface correspond to a pure, ringing oscillation that we can see but not hear. The earth's two fluid envelopes, the atmosphere and oceans, similarly permit music, on a very grand scale, that of our globe.

Two instruments that play prominent parts in the music of our atmosphere are the Jet Streams, the swift rivers of air that girdle the globe in midlatitudes, one in each hemisphere. The speed of a Jet Stream depends mainly on the temperature difference between the equator and pole. That difference induces motion. Hot air rises near the equator, where it is warm. To sustain this upward motion, air flows toward the equator at low elevations, in the opposite direction aloft. The rotation of the planet introduces a Coriolis force that deflects the poleward flowing air aloft toward the east, so that it becomes a Jet Stream. So swift is that eastward flowing river of air that flights from California to New York are significantly shorter than those in the opposite direction.

The Jet Streams amount to taut strings whose vibrations and

undulations produce the analogues of musical notes: cyclones, anticyclones, warm fronts, cold fronts, and tornadoes. Meteorologists scrutinizing weather maps that show how these features wander across the continent in the course of a few days are in effect perusing musical scores. The seasons provide the music with a rhythm. Although the beat is as rigid as that of a metronome — our planet orbits the sun with great regularity, once a year — the music is delightfully flexible. Exhilarating accelerandos can bring summer to a sudden and early end, but sometimes lazy ritardandos prolong that season, to such an extent that rhododendrons and other flowering trees bloom in October, as if summer were about to start anew. Such differences between one summer (or winter) and the next stem from a lively interplay between forced variability (induced by periodic variations in sunlight, the diurnal and seasonal cycles for example) and natural variability (the spontaneous music).

A clever experiment that sheds light on this interplay between forced and natural variability exploits the dependence of the Jet Streams on two main factors: the temperature difference between the equator and pole, and the rotation of the earth, which introduces a Coriolis force.[1] To simulate a Jet Stream in a laboratory, an annulus is filled with water (which represents the atmosphere) and is placed on a rotating table. Heating the outer cylinder of the annulus, the "equator," while cooling the inner cylinder, the "pole," creates a temperature difference that drives a Jet Stream — it causes the water in the annulus to flow from west to east. As the temperature difference between the equator and pole increases, so does the speed of the current in the annulus.

Figure 9.1 shows pictures of the annulus, as seen from above. The temperature difference between the equator and pole has its smallest value in the top left panel and then increases as we move to adjacent panels (as if reading words on a page). Each panel is therefore a snapshot in a progression from summer (when the temperature difference is small and the Jet Stream is relaxed) to winter (when the temperature difference is large). The pictures reveal astonishing changes in the behavior of the "Jet Stream" as it grows more and more intense. We see that orderly flow in con-

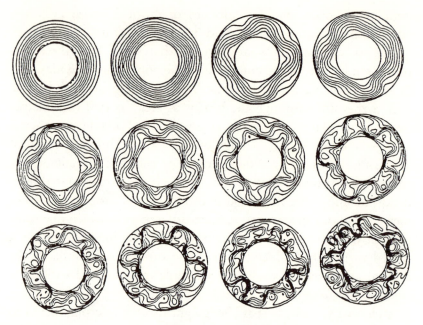

Figure 9.1. Flow patterns, at middle depth, of a fluid in an annulus when the inner and outer cylinders are at different temperatures. An increase in that difference can effect the transition from laminar to turbulent flow seen in this sequence of figures from Buzyna, Pfeffer, and Kung.[1]

centric circles gives way to spontaneous, modest meanders of the current and then to undulations that grow more and more chaotic. "Weather," associated with a meandering "Jet Stream," is absent from a world with a weak stream, is well behaved and predictable when that current is moderately strong, and becomes wild when the Jet Stream is intense.

The results in the figure explain why winter is not merely colder than summer but has different weather. In Chicago, for example, temperature changes in the course of a day can be dramatic in January—temperatures have been known to fall by tens of degrees within a matter of hours—but this does not happen in August, when the Jet Stream is relaxed and hardly meanders. The seasonal cycle, by causing a periodic intensification and relaxation of the Jet Stream, gives the spontaneous music of the atmosphere, the weather, a rhythm. In winter we tend to have severe storms that develop across the width of the continent; in summer weather is more local and is often associated with thunderstorms that develop late in the day. In technical terms, the forced, seasonal variability modulates the natural variability of the atmosphere. For a more poetic expression of the same idea we can turn to Shelley:

> O wild west wind, thou breath of autumn's being
> Thou, from whose unseen presence the leaves dead
> Are driven, like ghosts from an enchanter fleeing
> Percy Bysshe Shelley, "Ode to the West Wind"

Shelley is enchanted with the wildness and unpredictability of the wind, but he nonetheless recognizes that there is method to the madness. The method is evident in two predictable features: the wind is mainly from the west, and it appears mainly in autumn (rather than in the summer). Shelley distinguishes between these predictable, long-term tendencies and the unpredictable, wild, short-term changes. He in effect distinguishes between climate (forced variability) and weather (natural variability). Dictionaries give the impression that the weather of a region determines its climate, but these two separate phenomena, climate and weather, have a far more subtle relationship. Under certain conditions, weather can be completely absent, so that the winds are perfectly steady. (See the first panel of figure 9.1.) Hence it is possible for a planet to have a climate without having weather. If the winds on such a planet were to accelerate gradually, then the climate would change gradually. Once the winds become sufficiently intense, weather appears spontaneously, as the annulus experiments demonstrate.

The gradual intensification of a jet results, at a certain stage, in the appearance of spontaneous meanders. Such a transition to turbulent flow occurs when a controlling factor, the temperature difference between the equator and pole in this case, reaches a critical value. El Niño and La Niña influence that factor because they alternately warm and cool the eastern tropical Pacific. By doing so they alternately intensify and weaken the Jet Stream, and hence modulate weather. The Southern Oscillation between

	Natural Variability	*Forced Variability*
Atmosphere	Meanders of the Jet Stream (weather).	The seasonal cycle, forced by periodic variations in sunlight.
		Atmospheric aspects of El Niño and La Niña, in response to changes in ocean temperature.

El Niño and La Niña is therefore similar to the seasonal cycle; both cycles correspond to forced variability that modulates the natural variability of the atmosphere (see the "atmosphere" table). This perspective regards changes in sea surface temperature as externally imposed, as forcing the Southern Oscillation.

Not only the atmosphere but the ocean too has music, both forced and natural. The ocean takes its cues mainly from the winds blowing over its surface. A gentle breeze, like a violin bow that strokes a taut string, readily brings forth playful ripples on the ocean surface. The audible sound of a violin string is soothing when the pressure from the bow is light but becomes strident when the pressure is great. The ocean's music changes similarly, from ripples to choppy whitecaps to foaming, lashing waves as the gentle breeze grows stiff, becomes a strong wind, and then a gale that whips the ocean into a frenzy. This transition from innocent ripples to an angry ocean requires, not winds that fluctuate more and more wildly, but steady winds that grow in intensity.

Locally, winds excite waves on the ocean surface. On the vast scale of the entire ocean basin, the winds accomplish far more: they drive swift currents that include the majestic Kuroshio and Gulf Stream, and the maze of eastward and westward, surface and subsurface currents in low latitudes. As in the case of the atmospheric Jet Streams, so these rivers within the ocean spontaneously undulate and meander when they grow intense. During La Niña, when the trade winds and currents are particularly strong, photographs of sea surface temperature patterns in the eastern equatorial Pacific reveal waves that extend across thou-

	Natural Variability	*Forced Variability*
Ocean	Meanders of the Gulf Stream.	The tides, forced by the moon.
	Undulations of equatorial currents.	The seasonal cycle, forced by periodic variations in sunlight.
		Continually changing currents, in response to wind fluctuations associated with El Niño and La Niña.

sands of kilometers; those waves disappear when El Niño arrives and the winds and currents relax.[2] Hence the wind fluctuations induce the oceanic aspects of El Niño and La Niña, the variations in the intensity of the currents and in sea surface temperature patterns. This strictly oceanic perspective regards the wind fluctuations as externally imposed. They induce forced variability in the ocean, and that variability in turn modulates the natural music.

The "ocean" table, and the previous one for the atmosphere, list phenomena associated with forced and natural variability in the oceans and atmosphere. Note that the columns for forced variability appear to have inconsistencies. Some of the phenomena listed there, the tides and the seasonal cycle, have causes that ultimately are external to Earth. However, El Niño and La Niña also appear as forced variability even though their causes are not external to the planet. Furthermore, these phenomena involve a circular argument: the sea surface temperature patterns that force the atmospheric aspects of El Niño, including the winds, are in turn forced by those winds. Such interdependence suggests that the ocean and atmosphere are partners in a duet or dance.

To get things going, let us assume that the atmosphere takes the first step by providing the initial signal, a modest relaxation of the trade winds during La Niña. Those winds drive the warm tropical surface waters westward and cause the region of high temperatures, rising air, and heavy rainfall to be confined to the western Pacific. The ocean answers a slight initial weakening of those winds with an eastward movement of some of its warmest waters, thus expanding the region of high surface temperatures. This change in temperature induces a further weakening of the wind. Even more warm water now flows eastward, reinforcing the weakening of the winds. An initial, cautious retreat by the trades induces a tentative eastward step by the warm surface waters, which hastens the retreat, which emboldens the pursuit. . . . These interactions between the ocean and atmosphere cause the warm surface waters and humid air to surge across the tropical Pacific. Soon they are hugging the shores of Latin America. El Niño has arrived. Once El Niño is established, the stage is set for the dance partners to retrace their steps, to put La Niña in the spotlight. This new phase of the Southern Oscillation is an inversion, a mirror image, of the first part of this tango for ocean and atmosphere. A slight intensification of the winds drives some of the warm water westward, thus increasing the temperature difference between the eastern and western Pacific. That increase makes the winds stronger, which causes the temperature difference to become even bigger, thus enabling La Niña to evolve. As is usually the case, La Niña is the more lively partner. The oceanic currents associated with her grow so intense that they spontaneously start to undulate, adding arabesques to her fanciful sea surface temperature patterns.

Though intimately coupled, the ocean and atmosphere do not form a perfectly symmetrical pair. Whereas the atmosphere is quick and agile and responds nimbly to hints from the ocean, the ocean is ponderous and cumbersome and takes a long time to adjust to a change in the winds. The atmosphere responds to altered sea surface temperature patterns within a matter of days or weeks; the ocean has far more inertia and takes months to reach a new equilibrium. The state of the ocean at any time is not simply determined by the winds at that time, because the ocean is still adjusting to, and has a memory of, earlier winds, a memory in the form of waves below the ocean surface. They propagate along the thermocline, the interface that separates warm surface

	Natural Variability	*Forced Variability*
Ocean + Atmosphere	El Niño + La Niña	The recurrent Ice Ages, forced by periodic variations in sunlight. Global warming, in response to higher carbon dioxide levels.

waters from the cold water at depth, elevating it in some places, deepening it in others. These vertical displacements of the thermocline affect sea surface temperatures and amount to the hint that the ocean gives to the atmosphere that it is time to bring El Niño to an end and to start La Niña.[3]

The Southern Oscillation is natural and spontaneous; it is part of the music of our planet. If we wish the ocean, on its own, to produce that music, then the wind fluctuations associated with the Southern Oscillation need to be specified. The atmosphere on its own also has severe limitations. It can perform the dance only in response to specified sea surface temperature changes. That is why El Niño appears under the column "forced variability" in the two previous tables. He and La Niña emerge spontaneously when the ocean and atmosphere are intimately coupled, when the two are recognized as an inseparable unit. The tables therefore need a new entry in which El Niño and La Niña appear, not under "forced variability," but under "natural variability" (see the "ocean + atmosphere" table).

This new music too is constrained by superimposed beats, those of the recurrent Ice Ages associated with the very gradual variations in the intensity and distribution of sunlight over the millennia. Thus the properties of El Niño and La Niña — their frequency of occurrence for example — are subject to gradual changes. Scientists believe that El Niño was very different some 20,000 years ago, at the peak of the last Ice Age, and are hoping to find ancient corals from that time to check their speculations.

The English lady, after a few days in California, was bored with the monotonous weather. However, if she had stayed sufficiently long, she would have discovered that, in that part of the world too, weather can amount to interesting music. Some winters are drier and cooler than others; prolonged droughts alternate with periods of plentiful rainfall, even disastrous floods. If the weather of England were the music of a high-pitched violin or flute, then the longer-term climate fluctuations of California bring to mind a cello or bassoon. Other parts of the globe — India, where the monsoons fail occasionally, and the tropical Pacific, which has a Southern Oscillation between El Niño and La Niña — call for more instruments. Not a tuning fork, nor a single instrument capable of a tune, but only a huge symphony orchestra can do justice to the music of this planet.

TEN

A Marriage of the "Hard" and "Soft" Sciences

A winning wave, deserving note,
In the tempestuous petticoat, —
A careless shoe-string, in whose tie
I see a wild civility, —
Do more bewitch me, than when art
Is too precise in every part.
—Robert Herrick (1591–1674)

A physicist at a university in south-ern California starts his series of lectures with a dramatic demon-stration. He takes a large, heavy metal ball, suspended from the high ceiling by means of a taut cable, into his hands and steps backward until the ball just touches his nose. He then releases the ball, now a pendulum, so that it swings from him until it reaches the end of its arc. Next the ball starts to accelerate back toward the physicist. Everybody grows tense when he remains absolutely motionless, but to the relief of all, the ball comes to a halt just short of his nose.

This demonstration of confidence in the accuracy of the predic-tions that can be made on the basis of the laws of physics has an air of magic and never fails to impress. It is an excellent introduc-tion to the "hard" sciences, which focus on phenomena that can be studied by means of controlled, replicable experiments. The professor's demonstration can be repeated anywhere, anytime, because a pendulum always swings in exactly the same way. There's the rub. We admire and respect the accomplishments of the "hard" sciences, but too much emphasis on the timeless and universal can at times appear too mechanistic, "too precise in every part." Some of the Romantic poets of the nineteenth century felt that way about the results of Isaac Newton, the bearer of light for the Age of Reason. The painter Benjamin Haydon, after a festive literary dinner in his studio on December 28, 1817, wrote in his diary that

> Wordsworth was in fine cue, and we had a glorious set-to—on Homer, Shakespeare, Milton, and Virgil. Lamb got exceedingly merry and exquisitely witty. He then, in a strain of humor beyond description, abused me for putting Newton's head into my picture; "a fellow," said he, "who believed nothing unless it was as clear as the three sides of a triangle." And then he and Keats agreed that [Newton] had destroyed all the poetry of the rainbow by reducing it to its prismatic colours.

The Romantic poets preferred diversity and uniqueness over universality. Some turned their backs on the "hard" sciences and took a keen interest in the "soft" sciences because they delighted in nature's infinite variety of singular occurrences. For an appropriate introduction to the "soft" sciences—fields such as biology, ecology, and geology—the demonstration involving the pendulum should be modified as follows.

The professor teaches at a university in southern California, a region of frequent tremors and earthquakes. Suppose that a minor tremor, which causes no damage, occurs during the demonstration. The professor, a man of common sense despite his flair for miracles, promptly takes shelter under the nearest table. When he finally emerges, he could complain about the chance occurrence that ruined his beautiful demonstration. He would, however, be far wiser to regard that occurrence as a splendid opportunity to correct the common delusion that precise prediction is the hallmark of all science. (That misconception is evident when some people respond to inaccuracies in weather forecasts by declaring that weather prediction is "an art, not a science.") Many phenomena are so complex that precise predictions are impossible.

All natural phenomena, including the motion of the pendulum during and after the earthquake, are governed by natural laws and, in principle, are highly predictable. The classical Greeks were the first to propose that there are governing laws behind nature's great variety of particular occurrences. This idea was probably a projection onto the universe of the orderly life of the Greek polis, where a Greek citizen enjoyed considerable free-dom — from the whims of a ruler for example — even though his life was rigorously bound by impersonal laws. Nature would be less arbitrary and capricious than it seems to be if it too were governed by laws. Scientists have found that the laws of nature are far more absolute than civil laws. Whereas the latter can be disobeyed, with some inconvenience perhaps, those of nature must be obeyed; there is no choice. In principle, these laws permit precise predictions, but in practice, chance occurrences can pre-clude such predictions. (Charles Darwin, in a letter to Asa Gray, explained that all the universe runs by laws, "with the details, whether good or bad, left to the working out of what we may call chance.") Some unique phenomena, the evolution of the species *Homo sapiens*, for example, are so strongly influenced by contin-gencies that they cannot be explained as a simple consequence of the natural laws. Instead, explanations are in terms of a historical narrative, a reconstruction, after the fact, of what happened.

The validity of the explanations, as in the case of all science, depends on their testability. Some people are of the mistaken opinion that the most influential of all scientific explanations, Darwin's hypothesis that natural selection is the cause of evolu-tion, is not testable, that it amounts to a circular argument. They assert that natural selection, often described as "survival of the fittest," explains nothing, because "the fittest" are defined as those who survive. A careful consideration of the terms "sur-vival" and "fittest" makes it clear why this reasoning is falla-cious. "Fittest" refers to a trait that gives an organism an advan-tage in escaping from enemies, in finding food and mates, and so on. "Survival" refers to an entirely separate matter, namely birth and death rates. Organisms with a trait that gives them an advan-tage are more likely than those without the trait to have off-spring. Why should those offspring be as successful as their par-ents, especially if they live in a somewhat changed environment?

They obviously need to inherit the beneficial trait from their parents. Heredity is of vital importance to Darwin's theory, which therefore implies that a long lineage of similar organisms must all share the trait, perhaps with slight modifications from one generation to the next. Those modifications, which are random, can either increase or decrease the benefits an organism enjoys. Only those that provide an advantage persist over many generations. That is how certain traits evolve over many generations because of natural selection. This explanation for evolution, for the fact that species are mutable, may seem obvious — "How extremely stupid not to have thought of that!" remarked Thomas Huxley after reading Darwin's *Origin of Species* — but the explanation is by no means circular. It is testable. For example, the theory would be false if we could find a complex organ that could exist in no useful intermediate form. That is why natural historians focus on "the geological succession of organic beings, their distribution in past and present times, and their mutual affinities and homologies." Natural historians explain unique occurrences by searching for patterns in the rich and diverse record of detailed particulars in the past.

Weather forecasters too search for patterns in an unending sequence of unique occurrences. The particular arrangement of warm and cold fronts, cyclones and anticyclones, and so forth that at present covers the earth's surface — the weather at this moment — has never existed before. To anticipate how that arrangement will change over the next day or two, meteorologists carefully inspect a sequence of weather maps that depict the evolution of atmospheric conditions over the past few hours and days.

An alternative method for predicting the weather became available in the early 1950s, after the invention of the electronic computer. With this tool it is possible to predict the weather by solving the equations that express the laws governing atmospheric motion. This dramatic departure from previous practice seems to convert weather prediction from a "soft" into a "hard" science; objective computers solving precise equations replace subjective humans searching for changing patterns in complex maps. In principle, the computer can forecast the weather indefinitely into the future, provided that it is given perfect information about atmospheric conditions today, and provided that it is given instructions on how to solve the governing equations precisely. Are there, in reality, limits to the predictability of the weather?

The meteorologist Edward Lorenz addressed this matter in the 1950s and thus launched a new branch of science, popularly referred to as the study of chaos. This field deals with the complex behavior of seemingly simple systems that can be described in terms of precise mathematical equations.[1] Lorenz found that predicting the weather by means of a computer that solves equations may have the appearance of a "hard" science but that the results it produces are those of a "soft" science. Different scientists using the same precise equations to predict weather (or the same scientist repeating his exercise several times) produce, not the same forecast each time, but intriguingly different ones. The precise equations that "hard" scientists love and cherish apparently permit a diversity of solutions that will delight any "soft" scientist. The following simple example illustrates what Lorenz discovered.

Consider a wallet that falls from the pocket of a skier and slides down a ski slope. The path that the wallet travels is complicated because of the slope's innumerable little bumps and dips, shown in figure 10.1. These moguls, as they are known, can deflect the wallet sideways as it slides downhill. To calculate their effect, it is necessary to solve the equations that describe Newton's laws, subject to given "initial conditions." This term refers to the wallet's initial position, speed, and direction when it first falls onto the snow. In principle there should be no problem in determining where, at the bottom of the slope, the skier should look for his wallet, but in practice there are significant difficulties, as is evident in figure 10.2. That figure shows seven different trajectories for a wallet that, each time, starts with exactly the same speed and direction; only its initial position changes, by a mere 1 millimeter, each time. At first these differences are so slight that they have an insignificant effect on the path of the wallet. However, by the time the wallet has traveled a mere 60 meters downhill, some 180 feet, the differences have grown enormously. The paths continue to diverge rapidly thereafter. It follows that, in

Figure 10.1. Moguls on a ski slope. A skier is barely discernible in the upper part of the figure.[1]

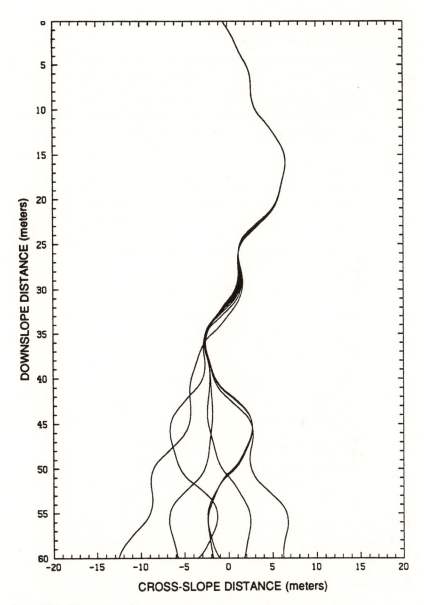

Figure 10.2. The paths of seven wallets, starting with identical velocity from points spaced at intervals of 1 mm along a west-east line.[1]

practice, it is impossible to predict where the skier will find his wallet if it slides too far downhill.

The scientific result that even phenomena that appear chaotic behave in strict accord with exact equations has an echo in Shelley's "Ode to the West Wind." He vividly describes the free-spirited "wild west wind," not in free-form poetry that gives him enormous freedom, but in a series of sonnets that have very rigid rules:

> O wild west wind, thou breath of autumn's being
> Thou, from whose unseen presence the leaves dead
> Are driven, like ghosts from an enchanter fleeing.
>
> Yellow, and black, and pale, and hectic red,
> Pestilence-stricken multitudes: O thou,
> Who chariotest to their dark wintry bed
>
> The winged seeds, where they lie cold and low,
> Each like a corpse within its grave, until
> Thine azure sister of the Spring shall blow
>
> Her clarion o'er the dreaming earth, and fill
> (Driving sweet buds like flocks to feed in air)
> With living hues and odours plain and hill:
>
> Wild spirit, which art moving everywhere;
> Destroyer and preserver; hear, O hear!
>> Percy Bysshe Shelley, "Ode to the West Wind"

This ode is composed of four more sonnets, for a total of five, each with strict rhythm (iambic pentameter) and rhyme. Note how the last words in the first and third lines, the second and fourth lines, and so on rhyme, thus providing a link from the one stanza to the next. These rigid rules create a sense of order and harmony. To portray a chaotic "wild west wind" whose random surges are followed by uneasy silence, Shelley departs from the sense of order imposed by a sonnet by writing long lines with random pauses that create a fluctuating tempo. (This becomes evident when the lines are read aloud.) He seems to anticipate the

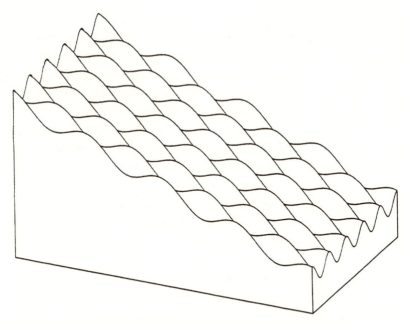

Figure 10.3. An oblique view of a section of the model ski slope.[1]

scientists who described turbulent phenomena, including wild winds, in terms of precise equations.

The results in figure 10.2 were obtained, not for an actual wallet sliding down a realistic slope such as the one in figure 10.1, but for an extremely idealized and simplified case. In the calculations, the moguls were the regular, idealized ones in figure 10.3, and the wallet, rather than being an object with a complicated shape, was an object shrunk to the size of a tiny pinhead! In spite of these considerable simplifications, the path of the wallet is predictable for only a very limited time! These results are humbling because, in principle, there is no problem. Finding an exact result is a matter of solving certain exact equations subject to initial conditions. Difficulties arise when we try to determine those initial conditions; innumerable chance occurrences make it impossible to determine exactly where the wallet falls onto the snow, certainly not to within a fraction of a millimeter. The tiny yet unavoidable errors introduced at the very beginning are the ones that ruin the predictions. Phenomena such as the sliding wallet are said to demonstrate extreme sensitivity to initial conditions. The prediction of the weather several days hence, which requires a description of the state of the atmosphere today, is similarly limited by inaccuracies in the description of that initial state. In picturesque terms, failure to take into account every butterfly that flaps its wings limits the predictability of weather.

Results from the study of "chaos" shed glaring light on the severe limitations of scientists, their restricted ability to predict something as apparently simple as the path of a wallet sliding down a ski slope. In principle there is no problem; accurate predictions are possible on the basis of the laws of physics. In practice there are immense difficulties. In the case of weather, those difficulties limit predictability to a week or so. If so, how is it possible to anticipate El Niño months in advance? Even worse, how can scientists hope to determine how our immensely complex planet will respond, decades hence, to perturbations such as an increase in the concentration of greenhouse gases?

We may have very limited ability to predict, on the first day of January, say, what the weather in Chicago will be on a specific day a week later, let alone on a specific day in July, but we can nonetheless forecast, in January, that an average day in July will be much warmer than an average day in January. This climate change between January and July, a consequence of the steady increase in the intensity of sunshine during that period, affects certain aspects of weather in a predictable manner. To explore the relation between weather and climate, we return to the simple pendulum.

Consider a pendulum in the form of a heavy metal ball suspended from the ceiling by a cable. Assume that the ball is so heavy that it gradually stretches the cable. A long pendulum is more ponderous than a short one, so the period of oscillation of the pendulum gradually increases as the cable stretches. Assume that this change, which is perfectly predictable provided that the ceiling is stationary, is from 1 to 10 seconds in the course of a month. If we happen to be in a region of earthquakes, and if random tremors occur continually, then the pendulum's period of oscillation will be unpredictable at a specific time but there will nonetheless be a discernible change over a month. Initially the times it takes for the completion of successive oscillations will be clustered around 1 second: 0.5 sec, 1.2 sec, 0.8 sec, 1.3 sec, 1.2 sec, and so on. But after a month the oscillations of the much longer pendulum will be clustered around 10 seconds: 9.0 sec, 12.1 sec, 11.1 sec, 8.5 sec, 11.5 sec, and so on. In other words, although it is impossible to predict how much the pendulum's period of oscillation will change from one oscillation to the next, it is possible to predict how much it will change on the average as the cable's length increases. The pendulum has a natural mode of oscillation, back and forth at a high frequency, and also has forced variability, the gradual lengthening of the cable. The latter affects the former.

Meanders of the Jet Streams, which are the origin of weather, correspond to natural modes of oscillation of the atmosphere. The ocean has its counterparts: meanders of the Gulf Stream in the Atlantic Ocean, for example. The ocean plus atmosphere, regarded as a unit with two inseparable components, has a similar phenomenon, the spontaneous Southern Oscillation between El Niño and La Niña. To ask why these various meanders and oscillations occur is equivalent to asking why a pendulum swings back and forth. These are all spontaneous, natural modes of oscillation. A gradual change in the length of a pendulum modulates its period. The corresponding factor that alters the period of the meanders of the Jet Streams and Gulf Stream is their intensity. A factor that can change the period of the Southern Oscillation is the depth of the oceanic thermocline that separates warm surface waters from cold, deep water; the deeper the thermocline, the longer the interval between visits from El Niño. From the 1960s to the 1980s the frequency of occurrence of El Niño decreased from approximately once every three years to once every five years. That change was associated with a deepening of the thermocline in the tropical Pacific Ocean.

Errors in the measurements of the initial location of the wallet in figure 10.2 limit our ability to predict where it will be at the bottom of the slope. The predictability of the weather tomorrow is limited similarly by errors in our description of the weather today. Both the wallet and the weather are said to have great sensitivity to initial conditions. Our ability to predict El Niño, a phenomenon far more regular than the weather, is limited for different reasons. El Niño is one phase of the continual Southern Oscillation, which can be regarded as a damped pendulum. If it were not for modest blows at random times, the pendulum would hang vertically and there would be no oscillation. When it is swinging, we can anticipate what the pendulum will do, but only until the next modest blow unexpectedly appears. Because of those random blows, droll, whimsical El Niño will always be able to surprise us.

All natural phenomena behave in accord with strict laws that express timeless, universal principles, but chance occurrences limit the degree to which those laws can be used to make accurate predictions. The precise location and timing of earthquakes are impossible to anticipate, but scientists can infer from the governing laws how the earth will move during an earthquake, thus making possible the construction of buildings that survive earthquakes. In the case of weather, and of El Niño, the governing laws allow only limited predictability of the evolution of complex patterns. This amounts to a marriage of the "hard' and "soft" sciences: the exact equations that permit limited predictability describe a world with infinite variety, even chaos. There is both method and madness, and no risk of art that is too precise in every part. These scientific results would have pleased Robert Herrick and probably the Romantic poets too.

ELEVEN

The Cloud

I am the daughter of earth and water,
And the nursling of the sky:
I pass through the pores of the ocean and shores;
I change, but I can not die.
For after the rain when with never a stain
The Pavilion of Heaven is bare,
And the winds and sunbeams with their convex gleams
Build up the blue dome of air,
I silently laugh at my own cenotaph,
And out of the caverns of rain,
Like a child from the womb, like a ghost from the tomb
I arise and unbuild it again.
—Percy Bysche Shelley (1792–1822), "The Cloud"

May I introduce myself? I have a great many names to match my infinity of shapes, but they are all variations on three basic themes.[1] On hot summer afternoons I often make my entrance as an innocent, fluffy cotton ball, but then I quickly grow into an ominous, tall tower that threatens with thunder and lightning. As such I am known as cumulus. As cirrus I am far less substantial, a wispy, transparent, windblown web of minute ice crystals at considerable altitudes. In this shape I lend the moon and sun halos when I pass in front of them:

> I bind the Sun's throne with a burning zone,
> And the Moon's with a girdle of pearl

As stratus I spread out horizontally to cover acre upon acre, forming an endless, sunlit prairie to those looking down on me from an airplane. Most of the time I am a combination of these three shapes, and by adopting Latin qualifiers — alto for high, nimbus for rain, and so on — I can be any of cirrostratus, cumulonimbus, stratocumulus. . . .

You seem surprised, and pleased, to learn that a few carefully chosen words can capture my innumerable shapes and mutations. You are delighted to find that, although I am ephemeral and evanescent, I am not necessarily a reminder that everything you cherish is fated to disappear, to recede into intangibility. The very opposite: "I change, but I can not die." I am a splendid example that nothing can arise out of nothing; nothing can be reduced to nothing. These insights, which your poets treasure, come from the world of science. Early in the nineteenth century the poet Samuel Coleridge explained that he attended scientific lectures to renew his stock of metaphors.[2] Shelley was particularly adept at expressing a scientific outlook in verse. To learn about me, I suggest that you read his poems.[3]

I appear, disappear, reappear, and readily change my shape because I am composed of a very remarkable chemical, water. Whenever I arise "like a ghost from the tomb," the invisible gas water vapor condenses into visible water droplets or ice crystals. When I disappear the reverse happens, and water reverts to its gaseous form. Many substances are capable of similar transformations; they too can change their phase from solid to liquid to gas as their temperature increases. Water is exceptional because it is the only substance that is present in all three phases in the relatively narrow range of temperatures encountered near the surface of the earth. In frigid ice, the water molecules are huddled together so tightly that they can only vibrate in place; no molecule is able to move past another. They gain freedom when ice turns into liquid water; molecules are then free to glide past each other. Being a gas adds exhilarating freedom, not only because water becomes invisible, but also because each of its molecules has complete freedom to move as it pleases. Freedom comes at a price. To transform 1 gram of ice at 0° C into water at 0° C requires 80 calories. The further addition of 1 calorie for every gram of water raises the temperature by 1° C. Such heating causes the water to boil at 100° C, at sea level. To convert the gram of liquid water at 100° C into the gas water vapor at the same temperature requires 540 more calories. This expense is recovered when the water molecules return to bondage. For example, when freely flowing liquid water turns into rigid ice, heat is released. Similarly, heat is released when water molecules in a gaseous state reestablish their bonds and condense to form water droplets. That happens when I make an appearance. I am therefore generous when I make an entrance—I provide energy to my surroundings. Where do I get this energy?

The sun is our planet's main source of energy. Since the oceans cover some 70 percent of the surface of the earth, they receive most of that energy, especially in the tropics, where sunlight is most intense. The winds prevent the tropical oceans from becoming excessively hot, by causing the evaporation of seawater. The oceans continuously gain heat from the sun and lose it through evaporation to the atmosphere. (It is for the same reason that a wet swimmer on a windy beach feels cold when the water droplets on her skin evaporate.) Once in the atmosphere, the water vapor is carried hither and thither by the winds. When the winds lift the moist air to higher elevations, the fall in temperature causes the vapor to condense into water droplets, and I appear. I celebrate the occasion by releasing heat. (That is why water vapor is said to have latent heat.) The heat makes me buoyant and enables me to grow bigger. I do this by entraining more air, converting its moisture into water droplets, and thus gaining even more heat for further growth. That is how a few cotton puffs that dot the sky on a warm summer afternoon can turn into a threatening, thundering cumulonimbus within a matter of hours, by voraciously entraining the surrounding air and using its moisture as a fuel. The winds take that fuel from the ocean, which loses heat through evaporation but recovers its energy from the sun. The winds therefore act as a lens that focuses the heat from sunlight that falls on the ocean onto the patch of sky where I appear. A hurricane is particularly effective at focusing heat in this manner. Its swirling winds gather water vapor from the ocean and deposit it in clouds that grow vertically, thus causing the winds to spiral inward faster and faster.

The intense trade winds that rush westward across the tropical Pacific during La Niña deposit their moisture in the cumulus towers that are confined to the region of very warm surface waters in the far western Pacific. When the warm water surges eastward during El Niño, evaporation from the ocean increases. The condensation of that water vapor into the clouds is accompanied by a huge increase in the release of latent heat. That is how El Niño gets the energy to cause strange weather globally.

Winds facilitate the transfer of water from the oceans into the air. What matters to me is not how much moisture the air actually has but how much additional moisture the air can absorb before it is saturated. I am sensitive to the relative humidity, the ratio of the moisture in the air to the moisture it would have if it were saturated. When the relative humidity is 100 percent, when the air is saturated, the further addition of water vapor leads to condensation. That is when I make my appearance — in the form of visible water droplets — and when your complaints about the humidity reach a crescendo. Your comfort depends on the relative humidity of the air. That is why you find New York intolerable in summer but claim to be comfortable in the dry desert air of Arizona, even when temperatures are high. "It's the humidity, not the heat," you explain. It is possible to be comfortable in spite of high temperatures if the relative humidity is low, because the dry air will readily accept moisture that evaporates from your skin. That evaporation keeps you cool and comfortable. Evaporation when temperatures are low will of course make you feel chilly. It may be the reason why, in winter, you feel cold in a room with low humidity.

One way to saturate dry air is to increase its concentration of water vapor. Another way, which does not require the release of more water vapor into the air, is to lower the temperature of the air. You do that when you fill a glass with water and ice on a humid day. The air in contact with the glass is cooled, and when its temperature is sufficiently low, the water vapor in the air spontaneously condenses onto the glass. (The liquid on the outside of the glass comes from the air, not from inside the glass.) It is bewildering to find that when I adopt this form, you consider me a nuisance. And yet you think I am delightful when, after a night of falling temperatures, I appear as morning dew on the plants in your garden!

When a layer of air above the ground, and not merely the ground itself, is cooled sufficiently during the night, then I appear in a lowly form, as haunting fog. Along the coasts of California and Oregon sea surface temperatures are lower near the shore than farther offshore. Air moving eastward toward the coast, toward San Francisco, is therefore cooled, becomes saturated, and I materialize. If the air were absolutely clean and free of dust, I could not make an appearance even at low temperatures, because impurities (nuclei) are essential for condensation to start. Thick fogs used to plague London, causing innumerable disasters such as people accidentally falling into the Thames River, but since the banning of open fires, fogs occur infrequently. (The fog appeared when winds brought water vapor to the soot-laden and hence nuclei-rich air.) The intentional addition of nuclei to clouds — seeding them — has been attempted in dry regions to enhance rainfall. Despite numerous attempts, it is still unclear whether such efforts are effective.

The rain I produce is sometimes gentle, sometimes a torrent. When I adopt my stratus form, I am capable only of drizzles of small raindrops. However, when I am a tall cumulus tower, water droplets can travel up and down inside me and, through accretion, can grow substantially in size before falling as rain. That is why, in my cumulus form, I can produce cloudbursts. When I am sufficiently tall for temperatures in my crown to be well below freezing, then droplets that reach those altitudes turn into ice. Repeated visits to my crown create hailstones that can grow to the size of golf balls before they are so heavy that they overcome the updrafts and fall to the ground. That is when

> I wield the flail of the lashing hail,
> And whiten the green plains under,

> And then again I dissolve it in rain,
>> And laugh as I pass in thunder,

In winter, in high latitudes, I do not have to be particularly tall for my temperatures to be below freezing. Under such conditions water droplets can turn into delicate ice crystals that sometimes stick together to form snowflakes in a variety of shapes. On such occasions

> I sift the snow on the mountains below,
>> And their great pines groan aghast;
> And all the night 'tis my pillow white,
>> While I sleep in the arms of the blast.

My presence requires continual evaporation, condensation, and precipitation. To what end? I do it for your benefit. If it were not for my ceaseless toil, you would find the surface of the earth intolerably hot. There is a simple way to estimate the degree to which evaporation from the surface of the earth keeps it cool. The globally averaged precipitation at the surface of the earth in the course of a year is approximately 1 meter, which must therefore be the amount of water that evaporates from the surface each year. The evaporation of all this water causes the earth's surface to lose heat to the tune of 83 watts per square meter, the equivalent of almost one-half of the sunlight that reaches the surface. To appreciate the enormous importance of this loss, consider what would happen in its absence. The temperature of the surface would have to increase until the heat that was radiated upward equaled that received from the sun. In that case the temperature, averaged over the globe, would be 67° C instead of the actual 15° C.

Evaporation, which feeds my growth, is but one way in which I cool the surface of the earth. I have another, more direct means of keeping the surface cool — by acting as a parasol that reflects sunlight. My white color, and the vast areas that I cover, make me so effective that some 30 percent of the sunlight incident on our planet is reflected back to space and thus fails to cause any heating. I provide compensation, however, because the water droplets and water vapor in clouds both provide a strong greenhouse effect. I therefore act as a blanket that keeps the surface of the earth warm. This is most evident on winter nights; temperatures are far higher in my presence than absence.

Depending on my shape or size, whether I am stratus or cumulus, I can either cool or heat the surface of the earth. No wonder scientists studying our climate find it difficult to cope with me. At present those scientists are trying to estimate how much global temperatures will rise because of the carbon dioxide humans inject into the atmosphere. That greenhouse gas is raising temperatures, thus increasing evaporation from the oceans. The warming is therefore magnified because water vapor too is a greenhouse gas. The higher temperatures can cause even more evaporation, even more water vapor in the air, and hence an even bigger greenhouse effect. Here we have the makings of a runaway greenhouse effect that stops only when all the water has evaporated from the oceans. That probably happened on the planet Venus. Earth, fortunately, is farther from the sun, so the greenhouse effect is smaller and temperatures are lower. As a result, evaporation came to a halt long before the oceans disappeared, when the air became saturated with water vapor and I appeared in my visible form.

You find me entertaining when I behave capriciously and enliven the sky, but in reality I am of crucial importance to the climate of your planet, sometimes heating it, sometimes cooling it. As stratus, I cover huge areas and reflect large amounts of sunlight, thus depriving the earth of heat. As cumulus, I have the shape of a tower and reflect relatively little sunlight, but I do have a strong greenhouse effect and hence I keep the earth's surface warm. Will I respond to the current increase in the atmospheric concentration of greenhouse gases by heating or by cooling this planet? Scientists are trying to answer this question by means of computer models that attempt to simulate me. Imagine computer models as whimsical as I am!

Part 4 | A Brief History of the Science

TWELVE

Predicting the Weather

I often say that when you can measure what you are speaking about, and express it in numbers, you know something about it; but when you cannot measure it, when you cannot express it in numbers, your knowledge is of a meagre and unsatisfactory kind.
—Lord Kelvin (1824–1907)

During the election campaigns of 1994, a congressman who proposed that the Department of Commerce be abolished was asked where he would get weather forecasts. "From television," the congressman replied, apparently unaware that the Weather Service is a branch of the Department of Commerce. Weather forecasts are so readily available from radio, television, and daily newspapers that most people take them for granted without wondering how they are produced. Until quite recently weather prediction was viewed as an augury, but today we regard those forecasts as reliable and important information. For example, during World War II an admiral in the U.S. Navy chose to ignore a warning that his fleet was in the path of a typhoon; 790 lives were lost.[1] Today officials order the evacuation of coastal cities and towns if the forecasts indicate that a hurricane is threatening. What were the key factors that contributed to the impressive advances in weather prediction? Are there any lessons for those attempting climate predictions, of El Niño or global warming for instance?

"Weather writes, erases and rewrites itself upon the sky with the endless fluidity of language; and it is with language that we have sought throughout history to apprehend it."[2] In the sixteenth century scientists developed a very special language for the study of natural phenomena such as weather, a language that uses mainly numbers. Lord Kelvin, in the epigraph at the beginning of this chapter, greatly admired this language, whose development required the invention of scientific instruments in order to make measurements, of atmospheric conditions for example. In the nineteenth century, measurements by groups of people, dispersed over large areas, were gathered in a few central locations for the purpose of producing weather maps and making weather forecasts. Today, weather prediction involves the continual acquisition of huge amounts of atmospheric data across the globe. Supercomputers at meteorological centers manage the flow of data and predict the weather by solving elaborate systems of equations that express the laws governing changes in atmospheric conditions. The language that meteorologists use when discussing the weather among themselves has become very specialized and is foreign to most people. However, when presenting forecasts to the public, the scientists use language (and maps) everyone understands. The flow of information is not simply in one direction, from the meteorologists to the public, because the forecasts generate such enormous interest that the scientists are obliged to respond. People are passionate about "their" weather and tend to be skeptical of the accuracy of impersonal forecasts made at distant weather prediction centers. They do not allow meteorologists to forget forecasts that are seriously in error. This critical attitude of the users of weather forecasts contributes directly to improvements in those forecasts. Meteorologists are reminded daily that, to justify the resources required for their science, they have to produce accurate predictions.

At first it may appear that, in the case of weather forecasting, the one main goal of science, providing practical benefits, has the upper hand over the other main goal, an understanding natural phenomena. Not at all. The history of weather forecasting is a wonderful example of how scientific progress results from a continual interplay between these two complementary goals of science. A weather forecast is usually presented as a sequence of maps showing how atmospheric conditions will change during the next several days. Each map has a unique arrangement of warm and cold fronts, cyclones and anticyclones, etcetera. The number of possible arrangements of those elements of a weather map is inexhaustible. It is as if we were dealing with a very complicated version of chess in which pawns, bishops, knights, and other pieces can be deployed in an unlimited number of ways. In the case of chess, one player tries to anticipate what the other will do. In the case of weather, the goal is to anticipate what nature will do next. The first step for any player is to become familiar with the pieces — the cyclones, fronts, etcetera — and the rules that govern their moves. This requires an understanding of the different weather phenomena and the interactions between those phenomena. Hence the final result, an improved forecast, even though it is the product of "big" science — of highly coordinated efforts involving huge numbers of people — depends critically on contributions from "small" science, on an improved understanding of natural phenomena. Each forecast is the beginning of a new game, which starts with an entirely new deployment of the pieces and then develops in an entirely novel manner. A loss — an inaccurate forecast — is analyzed in great detail. The players carefully balance the needs of "small" and "big" science in their efforts to improve the forecasts.

Weather prediction is a critical aspect of our affair with El Niño because it involves the acquisition of the huge data sets that enable us to describe climate fluctuations such as the Southern Oscillation and the associated variations in atmospheric conditions. Any science depends on the availability of measurements to test theories and hence depends on technological innovations such as the invention of instruments with which to make measurements. The instruments that permitted weather prediction to become a quantitative science include thermometers, invented late in the sixteenth century, and barometers, invented toward the middle of the seventeenth century. The one measures temperature, the other a less familiar parameter, atmospheric pressure. Evangelista Torricelli (1608–1647) designed a simple instrument to measure atmospheric pressure: he filled a tube with mercury and inverted it into a dish of mercury. The pressure of the atmosphere (in effect, its weight pushing down on the surface of the earth) was shown to be equivalent to the weight of a column of mercury 76 centimeters high.

To appreciate how enormous this pressure is, we have to recall the celebrated public experiment Otto von Guericke (1602–1686) conducted in Magdeburg, Germany in 1660. He demonstrated the magnitude of the pressure that the atmosphere exerts on all objects, including us, by pumping the air from a hollow sphere, made of two joined metal hemispheres, and then having two teams of eight horses try to pull the hemispheres apart. So great was the force with which the atmosphere pressed the two hemispheres together that the horses failed.

When the French philosopher Blaise Pascal (1632–1662) ascended a mountain carrying a barometer, he observed that atmospheric pressure decreases with elevation, thus establishing that, if we interpret this pressure as the weight of the column of air above us, then that column has a finite height. We in effect are living at the bottom of an ocean of air. In the 1780s the Montgolfier brothers initiated explorations of that "ocean" by means of hot-air balloon ascents. Those events generated enormous public interest; half the population of Paris, some four hundred thousand people, turned out to watch the very first manned ascent, from the Tuileries Gardens, in December 1783. (Shortly afterward the chemist Jacques Charles became the first man to see the sun set twice on the same day, by means of a balloon flight late in the day.) As in the case of manned space flights in the latter half of the twentieth century, so manned balloon flights drew enormous public attention but contributed only modestly to science. Detailed observations of the vertical structure of the atmosphere, by means of aircraft, rockets, and unmanned balloons, started in World War I.

Conditions at higher elevations interest us because they sometimes reach down to the surface, during storms for example. (Balloonists find that they can encounter the sleet, snow, and fierce winds of winter at higher elevations even on a balmy day in summer.) To anticipate tempests, it is useful to monitor changes in atmospheric conditions aloft. Fortunately, that can be done from the surface by means of pressure measurements. It is a common observation that a fall in pressure at the surface is usually associated with the appearance of overcast skies and foul weather, a rise in pressure with fair weather. What is the reason for this relationship? The first step toward an answer was the realization that weather, rather than being a strictly local phenomenon, involves continually changing spatial patterns over a large area. Documenting those patterns requires the coordinated measurements that are used to create weather maps.

In 1816, Heinreich Wilhelm Brandes (1777–1834), a professor of physics at the University of Breslau, prepared a sequence of daily weather maps for the year 1783. He used meteorological data from a set collected thrice daily, between 1781 and 1792, by thirty-nine volunteers in eighteen countries, mostly in Europe, under the auspices of the Meteorological Society of the Palatinate. (As early as 1654, Grand Duke Ferdinand II of Tuscany first sponsored the monitoring of atmospheric conditions over a large area—a dozen stations across northern Italy, the Alps, and into central Europe—but the project lasted only until 1667, when the organizing body, the Accademia del Cimento in Florence, disbanded. Over the course of the next century there were several similar projects, in different parts of Europe, to collect meteorological data over huge regions. None succeeded in generating much scientific interest in the weather.)[3] Brandes, by plotting lines of constant pressure (isobars) on his maps, and superimposing the intensity and direction of winds at different locations, discovered that severe storms are associated with centers of low pressure ringed by isobars. Tracking the path of a storm is therefore equivalent to tracking the movement of a center of low pressure. Storms are of course associated with clouds and rain. How are they related to the variations in pressure that are plotted on a map?

Brandes found that the winds seemed to blow toward the center of the storm, toward the center of low pressure. However, the measurements were so sparse that they permitted other interpretations. Thus a student of Brandes's, Heinrich Dove (1803–1879), who was director of the Preussische Meteorologische Institut in Berlin from 1849 to 1879, insisted that the prevailing winds were mainly from the southwest and the northeast. He proposed that shifts in the boundary where these radically different air currents clashed—the one bringing warm, moist air from lower latitudes, the other bringing cold, dry air from higher latitudes—caused changes in the weather.

Meteorologists in North America had different views on storms. James Pollard Espy (1785–1860), a very successful popular lecturer known as the "Storm King," attributed the formation of clouds to the rising of moist air. The fall in pressure with elevation causes the moist air to expand and cool. As a result, the water vapor in the air condenses into water droplets to form clouds. Espy identified the latent heat released when water vapor converts into liquid water as the "motive power" of storms. In the 1830s he started to speculate that, in times of drought, rainfall could be induced by starting fires at the surface of the earth, thus causing the air to rise. He proposed that timber lots be maintained between the Gulf of Mexico and the Great Lakes along the western frontier. Rain could then be provided on a regular basis— the same morning of every week, say—by setting some of the lots ablaze. Winds would carry the rain eastward, and fair weather would return in the wake of the storms. This proposal met with opposition, in part because southern politicians feared that the government would have too much control over their weather.[4]

Espy's explanation for storms implies that the surface winds converge onto the center of low pressure where the air is obliged to rise, thus creating clouds. William Redfield (1798–1857) had evidence to the contrary. In 1821, while journeying from western Massachusetts to his home in Connecticut several months after a violent storm had toppled trees throughout the region, Redfield noticed that the alignment of fallen trees indicated winds from the northeast at the beginning of his journey but winds from the opposite direction closer to home. He inferred that the storm had been circular, a whirlwind. In subsequent studies of numerous storms he corroborated these findings.

Different investigators had different views on the relation between winds and pressure in part because they were investigating different types of storms. Tornadoes, tropical hurricanes, and large-scale cyclones in midlatitudes have very different dimensions and hence involve different physical processes, but such distinctions would emerge only after more measurements had become available. The invention of the electric telegraph in the 1840s, an event of major importance to meteorology, led to a significant enhancement of measurement networks. This tool permitted the rapid collection of data in a central location and hence the preparation of maps for the prediction of weather. Shortly after telegraph wires linked Washington and Boston in 1845, both Espy and Redfield proposed that the Smithsonian Institution establish a meteorological department to study and forecast storms. In Europe, meteorologists made similar proposals. Support for an expansion of observational networks was consolidated when meteorologists argued that, by using the telegraph, a sudden storm that had devastated a joint British-French naval expedition near Balaklava in the Black Sea on November 14, 1854, could have been anticipated. (The storm had traveled across southern Europe over the course of several days.)

On August 1, 1861, the Royal Meteorological Office inaugurated daily forecasts for each of four districts comprising the United Kingdom. This happened even though meteorologists were still lacking a clear scientific understanding of storms. Many members of the British Association for the Advancement of Science were uneasy about this state of affairs and insisted on suspension of the forecasts. However, the public outcry was such that the daily forecasts were soon resumed. The predictions of storms and gales were proving valuable, especially to those involved in coastal shipping; the forecasts helped reduce the number of shipwrecks significantly. In the United States weather forecasting also suffered temporary setbacks, in part because the Civil War severed links with the South. Predictions were suspended after a fire destroyed the Smithsonian building but were resumed in 1870, when the responsibility for forecasts shifted to the telegraphic network of the Department of War.

During the second half of the nineteenth century the expansion of the observational network was accompanied by the establishment of national weather services, of professional meteorological societies, and of journals devoted to articles on atmospheric phenomena. Meteorology was becoming a separate academic discipline. As charts depicting both atmospheric pressure and winds became more reliable, several scientists realized independently that a reasonable theory for storms should incorporate aspects of both Redfield and Espy's ideas.[5] To maintain the upward motion of moist air at the center of a storm, as required by Espy's theory, the winds must have a tendency to move toward the center. However, Espy had failed to take into account that such motion would be affected by the rotation of the earth. Because of the Coriolis force, the winds must move mainly around the center of low pressure and spiral inward while doing so. Redfield and Espy had been emphasizing different aspects of this motion.

Toward the end of the nineteenth century meteorologists appeared to have a convincing explanation for storms, but then, as often happens in science, new measurements revealed flaws in the theories. Differences in the rates at which temperatures decrease with height above centers of low and high pressure (above cyclones and anticyclones) were not in accord with the theories. Furthermore, many storms seemed to have a source of energy other than the release of latent heat associated with the condensation of water vapor. A group of Norwegian meteorologists, referred to as the Bergen school, led by Vilhelm Bjerknes (1862–1951, father of Jacob Bjerknes), identified an additional source of energy that becomes available when a mass of warm air collides with a mass of cold air. Such an occurrence has some similarities with the apparatus shown in figure 12.1, in which a partition separates warm water from cold water. Removal of the partition in (a) results in tumultuous motion as the fluids rearrange themselves, until finally the less dense warm water floats on the denser cold water, as in (b). The Norwegians proposed that the collision of masses of warm and cold air along a front could similarly result in the release of (potential) energy and the development of storms. The ascent of warm air accompanied by the descent of cold air are the critical features of such storms. (Espy's hypothesis that the release of latent heat is the "motive power" of a storm is

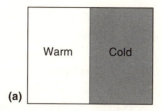
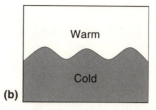

Figure 12.1. To illustrate energy conversions along an atmospheric front, consider cold and warm water separated by a vertical divider as in (a). Removal of the panel results in a rearrangement of the fluids as in (b).

less relevant to midlatitude cyclones than it is to tropical storms such as hurricanes.)

The upward motion of moist air that results in a storm can have its origin in high temperatures at the surface. The upward motion can also be induced by conditions aloft. For example, if conditions aloft favor the divergence of air from a certain region, then air from below will be sucked upward into that region. To anticipate such a storm, it is therefore necessary to prepare maps of conditions at the surface and also in the upper atmosphere, and to identify on those maps the warm and cold fronts that separate different air masses. These major innovations of the Bergen school of meteorologists required a significant increase in the number of measurements made in the upper atmosphere. Justification for such an increase came from the growth of commercial aviation after World War I and the attendant demand for information about atmospheric conditions aloft.

The weather prediction methods of the Bergen school at first met with resistance but, by the time of World War II, were widely adopted in the United States, mainly because of the efforts of the prominent meteorologist Carl Gustav Rossby (1898–1957).[6] World War II, and the need to predict the weather for military purposes, led to a huge increase in the number of meteorologists. They had ample opportunity to study conditions aloft because large number of sorties were flown in the upper atmosphere. They learned about the existence of the Jet Stream when planes flying at altitudes of 6 kilometers and more encountered very intense winds that caused several disasters. (Many bombers ran out of fuel because the wind speeds had been underestimated.)

After the war the invention of the electronic computer led to the introduction of an entirely new method for predicting the weather. The empirical method is to anticipate, on the basis of a series of weather maps for the past few days, how the patterns will change during the next few days. A more objective method is to exploit the physical laws that govern changes in atmospheric conditions. Already in 1925 L. F. Richardson proposed that the weather be predicted by solving the equations that express those governing laws.[7] He realized that, because the equations are horrendously complicated, predicting the weather would require simultaneous calculations by a large number of people. He whimsically described how sixty-four thousand people would perform their calculations in "a large hall like a theater, except that the circles and galleries go right round and through the space usually occupied by a large stage. The walls of this theater are painted to form a map of the globe. The ceiling represents the north polar regions, England is in the gallery, the tropics in the upper circle, Australia on the dress circle and the antarctic in the pit." In different parts of this room, different groups are making different calculations that need to be coordinated. To this end, "from the floor of the pit a tall pillar rises to half the height of the hall. It carries a large pulpit on its top. In this sits the man in charge of the whole theater. . . . he is like the conductor of an orchestra in which the instruments are slide-rules and calculating machines. But instead of waving a baton he turns a beam of rosy light on any region that is running ahead of the rest, and a beam of blue light upon those who are behind." Richardson also imagined that outside the hall "are playing fields, houses, mountains and lakes, for it is thought that those who compute the weather should breathe of it freely." Richardson personally made calculations to determine how the weather over Europe would evolve over a certain six-hour period. The results were a failure—there was no agreement with what happened in reality. Richardson believed the problem to be incorrect information about the winds at the time his forecast started.

In 1946 the mathematician von Neumann proposed that the newly invented electronic computer be used to bring to realization Richardson's dream of predicting the weather numerically. The equations that govern natural phenomena in the atmosphere were already known at the time, so that the availability of a computer would appear to make weather forecasting a straightforward exercise. To make that inference is equivalent to assuming that, once the rules of chess are known, that game becomes trivial. The equations for the atmosphere describe every possible phenomenon, from playful dust devils on the mesas of Arizona, to threatening hurricanes that can cover an entire state, to meanders of the Jet Streams that girdle the earth. No computer will ever be able to cope simultaneously with this enormous range of phenomena. To make progress, scientists have to filter out some of these phenomena by simplifying the governing equations. Which are the appropriate simplifications?

The secret of a good storyteller is knowing what to leave out. The meteorologists whom von Neumann assembled at the Institute for Advanced Studies in Princeton, New Jersey, in the early 1950s for the purpose of predicting the weather numerically faced a similar challenge. They had to simplify the equations that describe atmospheric conditions in order to isolate the very "essence" of weather, the basic processes responsible for storms, fronts, cyclones, etcetera. Jule Charney, the charismatic leader of this outstanding group—several members became prominent scientists subsequently—had derived the appropriate equations while a graduate student at the University of California in Los Angeles. He realized that weather will appear spontaneously in our atmosphere—a thin, spherical shell of gas on a rapidly rotating planet—when the temperature difference between the equator and poles becomes sufficiently large. That temperature difference determines the strength of the Jet Stream, which becomes more and more turbulent as it intensifies. Weather is a manifestation of that turbulence. This insight enabled Charney to identify the critical factors—they include the temperature difference between the equator and poles, and the rate at which the planet is spinning—that give rise to the development of large-scale weather phenomena such as cyclones and anticyclones. The simplified equations he derived are still being used by scientists to explain, in addition to weather, a great variety of phenomena that occur in rotating, stratified fluids such as the atmospheres of Earth and other planets, the oceans, and even the liquid core of the solid part of our planet. Ed Lorenz has described the derivation of those equations as the "greatest single achievement of 20th century dynamic meteorology." In Princeton, where Charney joined von Neumann, those equations were used in the first attempts to predict weather numerically.[8]

Richardson's failure at numerical weather prediction was at first a discouragement to his successors, but they quickly overcame several of the difficulties he had encountered. The problem of inaccurate wind information was circumvented when Charney developed a method that did not depend on that information. Furthermore, it soon emerged that the methods Richardson had used to solve his equations numerically were seriously flawed. New techniques were developed. So slow were the early computers, so long did it take to perform the necessary calculations, that keeping pace with the weather as it developed was a challenge. Over the next few decades the design of computers advanced rapidly, and the observational network for the atmosphere expanded enormously. The creation of an integrated global observational network to monitor weather was intimately related to the efforts of politicians to reconstruct a stable world order after World War II by promoting international cooperation in science and technology.[9] The need to adopt international standards of meteorological practice—for example, specific measurements with calibrated instruments taken at regular intervals and distributed internationally in a timely manner to facilitate weather forecasts—led, in 1951, to the creation of the World Meteorological Organization, an agency of the United Nations with headquarters in Geneva. (The predecessor of the WMO, the International Meteorological Organization, founded in 1873, was not an intergovernmental body, and its members did not act as official governmental representatives.)

The launching of *Sputnik* in 1957 made available a wonderful platform from which to observe the atmosphere from space. The attendant heightening of Cold War tensions led to more international programs to study the atmosphere. These culminated in the yearlong global atmospheric observational program — the Global Weather Experiment — that the World Meteorological Organization sponsored in 1979. The measurement platforms included several geostationary satellites (positioned over the equator at different longitudes) and polar orbiting satellites. Today those are components of an impressive global network of instruments that continuously monitors the atmosphere. Coping with this massive amount of data, which is now available on a routine basis, would have been impossible had it not been for the availability of more and more powerful computers.

Before he left Princeton in the mid-1950s, von Neumann urged the government of the United States to create a laboratory devoted to the use of computers to study weather and climate. That is how the Geophysical Fluid Dynamics Laboratory, now located in Princeton, came into existence. There, under the leadership of Joseph Smagorinsky, a small group of scientists including Syukuro Manabe and Kikuru Miyakoda used the increase in computer resources to take into account some of the supposedly secondary effects that the original Princeton group of weather forecasters had neglected. For example, they introduced a hydrological cycle, thus enabling them to predict rain, snow, or sleet; they expanded the domain being considered, from the northeastern United States to the Northern Hemisphere and then to the entire planet; and they pioneered the use of computer models of the atmosphere to cope with measurements of atmospheric conditions. The latter activity is important because some of the measurements are flawed. A ship in the Atlantic Ocean that makes an error in transmitting its location — a digit in a number that designates longitude may be wrong — could appear to be reporting on the temperature, winds, etcetera, in the middle of the Sahara desert. Such errors are easy to spot. More problematic are temperatures that are consistently too high or low at certain stations be-cause of faulty instruments. The computer can help identify and correct such problems in the following manner. If the computer models were absolutely perfect, then there would be no need for measurements, and if some were made anyway, then the computer would identify those that were incorrect. However, not only the measurements but the computer models too are flawed. It is then useful to rely on the strategy we adopt when we draw a straight line through several points. We need only two points to draw a straight line, but if the measurements that provide those points have uncertainties, we make additional, redundant measurements and draw a line that is the best fit to all the points. The computer model similarly provides us with all the information we need about temperatures around the globe, say, but that information has uncertainties. We therefore incorporate additional, redundant information in the form of measurements and use the computer to find the best fit to all these data. The methods for doing this "data assimilation" into models have become so sophisticated that forecasting centers are able to identify errant instruments in remote parts of the globe. Since the "assimilated data" provide us with the best description of atmospheric conditions in the past, those data sets are routinely used by meteorologists to analyze atmospheric processes and phenomena and to study long-term climate fluctuations and climate changes.

The development of computer models of the atmosphere started in Princeton in the 1950s and shortly afterward at the University of California in Los Angeles, at the National Center for Atmospheric Research in Boulder, Colorado, and at the Meteorological Office in Bracknell, United Kingdom. This activity has flourished, and today, at approximately a few dozen institutions worldwide, the most advanced supercomputers available are being used to develop models not only of weather but of climate too. Over the past few decades, an enormous international effort that required the commitment, skill, and imagination of a large number of people has brought Richardson's dream to reality.[10] Weather prediction has become a highly successful and specialized activity that integrates the contributions of instrument makers, rocket and

space scientists, computer engineers, mathematicians, and meteorologists across the globe. They have capitalized on advances in
the design of computers and in the design of instruments for observing the atmosphere from space. They have coordinated large
international research programs to observe the atmosphere in detail over prolonged periods. The product of their joint efforts is
the daily weather forecast. It involves the assimilation of measurements into a realistic model and thus generates an accurate
description of the continually changing atmospheric conditions.
Today everybody uses the products from the computer models in
a variety of applications, especially in studies of changes in atmospheric conditions associated with long-term climate fluctuations
such as the Southern Oscillation.

Weather is a global phenomenon, and in principle we need
only one center to produce a weather forecast for the entire
planet. In western Europe, several nations jointly sponsor the European Center for Medium Range Weather Forecasting, which is
located in Reading, England. Each of those nations continues to
operate its own weather services. For economic and security reasons, nearly every country has a weather service. The United
States has three forecasting centers: one for the air force, one for
the navy, and a civilian one, the National Atmospheric and Oceanic Administration (NOAA) in the Department of Commerce.
Commercial weather services, which provide the forecasts available on television, for example, start with the NOAA product
and tailor it to suit the specific needs of certain users. These various forecasting centers have both an operational and a research
branch, thus ensuring close interaction between those who try to
understand weather and those who predict it. The results of the
researchers are not proprietary and are therefore published, after
reviews, in the scientific literature. The entire research community
in effect participates in the improvement of weather forecasts.

The success of weather forecasting is an excellent demonstration of how, in science, progress depends on the availability of
measurements to test theories. In the atmospheric sciences, and
more generally the earth sciences, the measurements have to be
plentiful because the phenomena of interest do not repeat them-

selves (except for the tides) and do not lend themselves to replicable laboratory experiments. For example, the pattern of weather phenomena — fronts, cyclones, etcetera — that cover the globe at any moment has never occurred before and will not appear again. We learn about such phenomena, and become skilled in predicting them, by observing many members of the family. Weather forecasts have become reliable because the phenomena involved change rapidly; meteorologists are tested daily.

In the same way that the pattern of weather phenomena at any moment is unique, so each El Niño is unique. Up to now, scientists have observed relatively few of those phenomena because they occur infrequently, every few years. Forecasts of El Niño are bound to improve over the next few decades as we become familiar with more members of the family. When we turn our attention to the study of long-term changes in the properties of El Niño, and to global warming over the next several decades, we face a dilemma. The instrumental records available to address these issues are far too short to test theories. We will have adequate records many decades and centuries hence, but we need answers much sooner. To find tests for the models we are obliged to investigate Earth's past climates. (We return to this topic in chapter 16.)

THIRTEEN

Investigating the Atmospheric Circulation

Men in the north are tall in stature and fair in body because the north wind dries out and cools the land, and closes the pores of the body which then retains heat better. The south wind, which is hot and moist, has the opposite effect so that men of the southern lands are different from those in the north in stature and appearance. They are not so bold, nor so wrathful.

—Bartholomew, a Franciscan friar of the thirteenth century

Musings on climate often have interesting scientific implications. For example, Friar Bartholomew's observation that the south wind transports warm, moist air from lower to higher latitudes raises important questions. The south wind is always warm and must therefore have an inexhaustible supply of heat. The source of energy is obvious: the sun continually provides warmth to low latitudes. The south wind also carries air from lower to higher latitudes. Is there a continual creation of air in the tropics to maintain a steady supply for the south wind? No one believes that to be the case. Rather, we expect that, if south winds prevail in one region, then there must be north winds elsewhere, to return the air (possibly depleted of its heat and moisture) to lower latitudes. This means that the atmosphere has a circulation that connects different parts of the globe, distributing heat and moisture and thus creating different climatic zones.

The winds, by moving warm air from low to high latitudes, prevent the tropics from becoming excessively hot, polar regions from becoming excessively cold. The winds thus influence temperatures at the surface of the earth, but that influence is subject to constraints. For example, globally averaged temperatures must be consistent with the limited amount of sunlight available for heating the planet. There are similarly constraints on the distribution of moisture. The winds are free to bring different amounts of precipitation to deserts and jungles, provided that the net amount of rain that falls on the surface of the earth, in a state of equilibrium, equals the amount of water that evaporates from the surface, mostly the oceans.

Over the past few centuries, studies of the two aspects of Earth's climate — of the global constraints that determine, for example, the average temperature of the surface of the planet, and of the atmospheric circulation that distributes heat and moisture to create different climatic zones — have proceeded in parallel. Only with the availability of computer models of the atmosphere could the two aspects be merged. Only in the 1960s was it possible to start developing comprehensive, quantitative models of the atmospheric circulation. This provided atmospheric scientists with a powerful new tool they could use for controlled experiments, for example. Consider the changes in atmospheric conditions associated with the Southern Oscillation, changes that are usually attributed to fluctuations in the temperature of the surface waters of the eastern tropical Pacific Ocean. A realistic model of the atmosphere is a tool for testing that hypothesis. It can be used in a series of "numerical experiments" to determine the atmospheric response to different hypothetical surface temperature patterns. In the 1980s such experiments confirmed that simulation of the Southern Oscillation during a certain period is possible only if the sea surface temperatures actually observed during that period are specified. This is a very important result because it implies that models capable of predicting the weather a few days hence can also predict, months in advance, whether Peru and California are likely to experience exceptionally heavy rains, provided that we know how sea surface temperatures will change during the coming months. To deal with sea surface temperature variations, we need the oceanic counterpart of the atmospheric model to reproduce the oceans' response to the fluctuating winds. When the appropriate atmospheric and oceanic models are coupled together, we will presumably have a tool capable of forecasting phenomena such as El Niño. We return to this matter in a later chapter and confine our attention here to the development of models of the atmospheric circulation.[1]

The globally averaged temperature at the surface of the earth depends not only on the sunlight incident on our planet, minus the fraction reflected by clouds and other bright surfaces, but also on the composition of the atmosphere, on its concentration of greenhouse gases. In 1824 Jean-Baptiste Fourier explained that the atmosphere acts as a blanket that keeps the earth's surface warm — the atmosphere provides a greenhouse effect — because it is transparent to sunlight from above but is opaque to the "obscure radiation" (infrared radiation) from the earth's surface. Subsequent studies found that the gases responsible for the greenhouse effect are not the most abundant ones, oxygen and nitrogen, but are gases present in very small amounts. Particularly important are carbon dioxide and water vapor. (The major constituents of the atmosphere were already known by the end of the eighteenth century.) So large is the effect of these greenhouse gases that Earth's climate is very sensitive to small changes in their concentration. In a famous paper published in 1896, Svante Arrhenius argued that, in the wake of the industrial revolution, the atmospheric concentration of carbon dioxide would rise and would result in global warming. At first many scientists thought that the oceanic absorption of that gas would prevent its accumulation in the atmosphere, but in the 1960s Charles Keeling's direct measurements established that the atmospheric concentration of carbon dioxide is rising rapidly.[2] This finding renewed interest in global warming and motivated scientists to refine Arrhenius's calculations by taking into account that the heating of the earth's surface by sunlight is balanced, not only by the infrared radiation that the surface emits, but also by the atmospheric circulation that transports heat and moisture vertically (by means of convection) and also horizontally. Atmospheric winds, which are prominent features of that circulation, have generated discussions since the days of Fra Bartholomew and even earlier. Interest in the global pattern of winds grew significantly when navigators started their explorations in the fifteenth century.

The early explorers sailed from midlatitudes, where wildly fluctuating westerly winds prevail, and were therefore surprised to find that the prevailing winds in the tropics, the trades, are easterly and remarkably steady. Their reports of these winds generated much discussion. Some natural philosophers proposed that the easterly winds are the exhalations from the sargassum weeds in the subtropical oceans. Others suggested that the warm tropical air moves westward because of its lightness — the air is unable to keep up with the earth's surface in its rotation from west to east. In 1735 a London barrister, George Hadley, proposed an explanation that, by and large, is correct. He identified the temperature difference between the equator and poles as the main driving force for the atmospheric circulation. He realized that if our planet did not rotate, then warm air would rise spontaneously in low latitudes, where surface temperatures are at a maximum, and would sink back to the surface in the colder, higher latitudes. At the surface, air would flow toward the equator; aloft, it would flow poleward. The rotation of the earth alters this pattern radically. If the air retains its original eastward speed as it moves toward the equator in the lower atmosphere, then the surface of the earth below the air will be moving eastward at a faster and faster pace the lower the latitude. (On a rotating sphere, eastward motion is most rapid at the equator.) To an observer fixed on the surface of the earth, the winds will appear to be the easterly trades. Hadley estimated that the speed of the winds near the equator would be far greater than any speeds ever observed and assumed that frictional drag at the surface of the earth slows the winds down.

These arguments explain the westward trade winds of the tropics, except that the drag required to slow down the winds introduces another problem: the drag would also slow the earth down until, ultimately, it would not be rotating at all. This has obviously not happened, which means that the drag of the trade winds in the tropics is opposed by a drag in the opposite sense in the extratropics, that of the westerly winds in midlatitudes. But why are there westerly winds in the extratropics? Hadley proposed that the air aloft, which blows from the equator toward the poles, starts its journey with a high eastward speed and retains that speed as it moves poleward. Hence, relative to the surface of the earth, the winds become more and more westerly with increasing latitude. The air also cools as it moves poleward and sinks, so westerly winds appear at the surface in higher latitudes.

In the nineteenth century, refinement of the ideas in Hadley's remarkable paper led to the picture illustrated in figure 13.1, which explains not only the distribution of winds but also the location of regions of heavy rainfall near the equator and in midlatitudes, where moist air rises, and of deserts in the horse latitudes of the subtropics, where dry air subsides. This sketch is a good example of the idealizations scientists introduce in their attempts to cope with complex reality. It is reminiscent of the story about a physicist who was asked to advise a dairy farmer on how to increase the productivity of his cows. At their first meeting the physicist drew a circle on a sheet of paper and said, "Let us consider a spherical cow." Hadley advanced our understanding of Earth's climate enormously by considering a spherical planet without hills, mountains, valleys, or canyons; with no difference between land and sea; without any variations in longitude; and without time-dependent phenomena such as the weather. The atmospheric circulation on such a hypothetical planet, that shown in figure 13.1, resembles that of the earth in many respects, but there are discrepancies between those results and the climate we observe on earth. To remove them, the idealizations have to be relaxed.

In Hadley's circulation, the air that rises at the equator and then, aloft, flows poleward, becomes more and more westerly with increasing latitude. Toward the end of the nineteenth century some scientists estimated what those velocities ought to be, on the basis of theoretical considerations, and found them to be extremely high in midlatitudes, far higher than any observed speeds. Attempts to explain this discrepancy between theory and

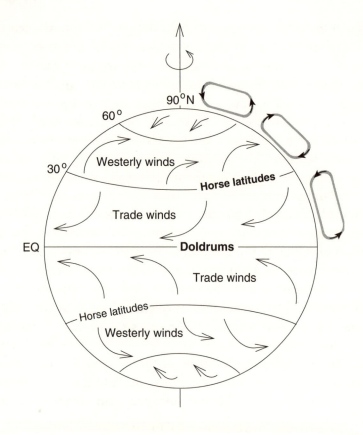

Figure 13.1. The atmospheric circulation on a water-covered globe. The north-east and southeast trade winds in the tropics converge onto the doldrums, rainy regions where moist air rises. In the subtropics, the easterly and westerly winds diverge from the horse latitudes, sunny, dry regions over which dry air subsides. The easterly winds in polar regions, and westerly winds of midlatitudes, converge onto rainy regions in the neighborhood of 60° N and S latitude. In reality, the presence of continents complicates the picture considerably.

observations gradually led to the realization that excessively high wind speeds are inhibited when the air's motion becomes turbulent. That turbulence manifests itself as the chaotic motion associated with weather. This means that the atmospheric circulation responsible for our climate is also responsible for our weather, an idea that is implicit in the arguments of the Norwegian meteorologist V. Bjerknes and his collaborators, who explained how cyclones can be regarded as disturbances that depend on a preexisting flow pattern, namely, the atmospheric circulation.

Weather depends on the atmospheric circulation that determines climate. (A planet with a variety of climatic zones but no weather — no day-to-day fluctuations in atmospheric conditions — is entirely possible. See chapter 9.) On our planet weather not only depends on the climate but also influences the climate. We complain bitterly about severe winter storms, but in reality those storms increase the habitability of the earth — improve its climate — by warming up high latitudes, cooling off low latitudes. That is what the meanders of the Jet Stream shown in figure 13.2 accomplish. At times, those meanders are so energetic that they fold back on themselves, creating cyclones and other phenomena associated with weather. In the figure, the meanders are seen to effect an equatorward flow of cold air and a poleward flow of warm air. The net result is a significant transport of heat from low to high latitudes.

Weather phenomena, and the atmospheric aspects of climate phenomena, all obey the same laws of nature, so computer models that use those laws to predict weather can, in principle, be used to study the atmospheric circulation associated with climate. The predictability of the weather is of course limited to a few days, but what would the results be like if the calculations were continued for a prolonged period of many months? At the Institute for Advanced Studies in Princeton, New Jersey, in the early 1950s, Norman Phillips made such an "infinite" forecast and obtained very encouraging results.[3] The experiment reproduced certain features of the atmospheric circulation: surface easterlies in low and high latitudes with westerlies in between. Aloft, a Jet Stream appeared. With this success Phillips launched the development of computer models of Earth's climate.

Because of the limited capabilities of computers, simulations of changes in global weather patterns over the course of a few days require approaches that differ from simulations of changes in the atmospheric circulation over the course of many years or centuries. In the case of day-to-day weather simulations the limited

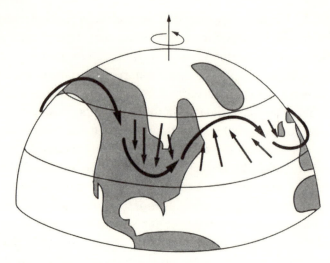

Figure 13.2. How meanders of the Jet Stream allow cold air to move southward while warm air flows northward.

computer resources are used to cope with regional aspects of weather that involve phenomena with relatively small dimensions — for example warm and cold fronts, the sharp boundaries between adjacent regions with very different atmospheric conditions. Atmospheric circulation models sacrifice that level of detail in order to simulate changes over much longer periods. The goal in climate studies is not the realistic reproduction of each storm but the realistic simulation of the statistical effect of a large number of storms. Hadley originally considered a highly idealized, perfectly spherical Earth and discussed it qualitatively, without making allowance for the effects of weather on climate. The models today are quantitative and take into account not only weather but also the presence of continents and oceans with different thermal properties, the presence of mountains and glaciers, the hydrological cycle that involves a variety of clouds, the greenhouse effect associated with numerous different gases, and many other phenomena. Such computer models of the atmospheric circulation are now capable of realistic simulations of the observed state of the atmosphere.[4] These simulations require that information concerning the composition of the atmosphere, and conditions at the surface of the earth, be provided to the models. A model of the atmosphere, by itself, is therefore only a component of a climate model, but it is nonetheless a powerful tool for studying, for example, climate fluctuations induced by changes in the land, ice, and ocean surfaces below the atmosphere.

A climate fluctuation that has received considerable scientific attention since at least the latter half of the nineteenth century is the sporadic failure of the monsoons over India; poor rains over the subcontinent used to contribute to famines. It is reasonable to start by investigating the reasons for the monsoons.[5] When the Indian subcontinent is hot during the summer, the rain-bearing southwest monsoons prevail; in winter the winds reverse direction, and there is essentially no rain. This description makes it sound as if the monsoons amount to a gigantic land-sea breeze, but the phenomenon is far more complex:

> I believe very few educated people would have any difficulties in giving an answer to the question—what is the cause of the monsoon? They would refer to the high temperature over the land compared with that over the surrounding seas; would speak of ascending currents of air causing an in-draft of sea-air towards the interior of the country. It is only when one points out that India is much hotter in May before the monsoon sets in than in July when it is at its heights—or draws attention to the fact that the hottest part of India—the northwest gets no rain at all during the monsoon—or shows by statistics that the average temperature is much greater in years of bad rains than in years of good rains, that they begin to doubt whether they know the real cause of the monsoon.[6]

What is the cause of the monsoons and, even more important, the cause of year-to-year variations in the intensity of the monsoons? Toward the end of the nineteenth century many people believed that the appearance of dark spots on the surface of the sun strongly affected atmospheric conditions and, in particular, the monsoons. Sir Norman Lockyer, the influential editor of *Nature*, declared that "the riddle of the probable times of occurrence of Indian famines has now been read, and they can be for the future accurately predicted, though not yet in various regions."[7] Blandford, the first director of the India Meteorological Department, created in 1875, believed otherwise. He was asked to investigate the causes of the drought that had contributed to the famine of 1877 and demonstrated that variations in the snow cover over the Himalayas had strongly influenced the monsoons. He used that result to anticipate droughts in the early 1890s, but his methods failed a few years later when India experienced yet another drought and famine. Renewed efforts to predict the monsoons brought Gilbert Walker to India early in the twentieth century.

Walker realized that failures of the monsoons are not simply local phenomena but are aspects of global climate fluctuations. His analyses of records from stations across the globe revealed that

> there is a swaying of press. [pressure] on a big scale backwards and forwards between the Pacific Ocean and the Indian Ocean, and there are swayings, on a much smaller scale, between the Azores and Iceland, and between the areas of high and low press. [pressure] in the N. Pacific.[8]

Walker named the Pacific "swaying" the Southern Oscillation and established that it is correlated with fluctuations in the rainfall, intensity of the winds, and a few other variables at a number of stations remote from each other. He was unable to provide a physical explanation for these surprising correlations. Furthermore, although he found that the warm phase of the Southern Oscillation tends to be associated with poor rains over India, he was unable to convert his correlations into a scheme for predicting the monsoons. After he left India and returned to England, interest in the Southern Oscillation waned for several decades but revived after World War II, especially after Bjerknes in the 1960s identified sea surface temperature variations in the eastern equatorial Pacific as the critical factor.[9] A rapid increase in the availability of atmospheric data, mainly for the purpose of weather prediction, benefited climate studies too. Some people had been skeptical of Walker's results concerning the Southern Oscillation because the paucity of the data he had analyzed implied a low statistical reliability. By the end of the 1970s, analyses of new data sets far more extensive than those available to Walker vindicated his results. In addition, Rasmussen and Carpenter were able to describe the development of a canonical warm El Niño—the warm phase of the Southern Oscillation—in detail.[10] By that time computer models proved capable of reproducing the atmospheric aspects of the Southern Oscillation, given the changes in sea surface temperature patterns. These studies confirmed Bjerknes's proposal that the atmospheric response to those patterns explain Walker's correlations between different variables in different places.

The changes in sea surface temperature associated with the Southern Oscillation are confined mainly to the eastern tropical Pacific and have a direct effect on atmospheric conditions in the tropical Pacific.[11] Northern Peru and Ecuador are almost certain to have heavy rains during El Niño. Conditions in regions farther away are more problematic because those conditions depend on far more than the temperature of the eastern tropical Pacific. Wallace and Gutzler established that peak El Niño conditions during the northern winter increase the likelihood of mild winters over central Canada and the northeastern United States, but those regions can have mild winters even in the absence of El Niño.[12] There is also an increased probability of heavy rains over California and the Gulf states, of decreased hurricane activity over the Atlantic, and of poor rains over India and southeastern Africa during El Niño. However, these statistical relations are subject to gradual changes. For example, in recent decades El Niño has had little impact on India.[13] Both India and Zimbabwe had normal rainfall during the intense El Niño of 1997. There are clearly factors that can overcome the effect of El Niño on rainfall in India and Zimbabwe, but as yet they have not been identified.

The change in the monsoons from one year to the next remains a major challenge because the change depends on a complex interplay between local and global factors. For example, on a water-covered globe, the rain belt in low latitudes onto which the surface winds converge would move back and forth across the equator as part of the seasonal cycle. The air that rises in the rain belt sinks over the adjacent regions, which therefore are deserts as in figure 13.1. The Sahara is presumably an example of such a desert in low latitudes. Even though surface temperatures in the Sahara are high, the air cannot rise to altitudes sufficient for convective clouds to form, because of the subsiding air aloft. The Sahara is dry because nonlocal factors overwhelm local factors. In India, because of its different geography, local factors can prevail, to a degree that varies from year to year. The task of predicting the monsoons, assigned to Walker more than a century ago, is proving very difficult. We are not yet able to weigh, in a given year, the relative importance of the many different factors that influence rainfall over India.

Since Hadley's pioneering study, we have made enormous progress in describing and explaining the circulation of the atmosphere. We have furthermore developed computer models capable of realistic simulations of the main features of the atmospheric circulation. An atmospheric model is but one component of a climate model because it requires that conditions at the land, ice, and ocean surfaces below the atmosphere be specified. A strictly atmospheric model is therefore inadequate if we wish to investigate how climate will change in response to, say, a doubling of the atmospheric concentration of carbon dioxide, because such a doubling will increase sea surface temperatures. To calculate that aspect of the response requires a model of the ocean. Matters become even more complicated when we ask questions, not about the effect of a fixed (permanent) concentration of carbon dioxide in the atmosphere, but about the response to a certain amount of carbon dioxide that is continuously injected into the atmosphere. Even if we emit into the atmosphere as much carbon dioxide as it already has, the result will not be a doubling of carbon dioxide levels, because the oceans and the plants will absorb some of that gas. To deal with this problem we have to investigate the carbon cycle — the cycling of carbon between our planet's various components. That task requires that a model of the atmospheric circulation be coupled to models of the ocean, the biosphere, the ice volumes. . . . Climate is an immensely complex phenomenon whose simulation requires models of the different components of our planet, all coupled together. We next turn our attention to one of those components, the oceans.

FOURTEEN

Exploring the Oceans

There is, one knows not what sweet mystery about this sea, whose gently awful stirrings seem to speak of some hidden soul beneath; like those fabled undulations of the Ephesian sod over the buried Evangelist St. John. And meet it is, that over these sea-pastures, wide-rolling watery prairies and Potter's Field of all four continents, the waves should rise and fall, and ebb and flow unceasingly.

—Herman Melville (1819–1891), *Moby-Dick*

In 1751 Henry Ellis, the captain of a ship off the sweltering coast of West Africa, was pleased to discover that, at a surprisingly short distance below the ocean surface, temperatures are so low that the water from those depths "supplied our cold bath, and cooled our wines . . . which is vastly agreeable to us in this burning climate."[1] In low latitudes the very warm surface waters are confined to a shallow layer that floats on the much colder abyssal ocean. So shallow is that surface layer that undulations and "gently awful stirrings" of the thermocline, the interface that separates the warm water from the cold, alternately expose cold water to the surface during La Niña and hide it during El Niño. These phenomena are of such importance to us that, in the late 1980s, we started maintaining an array of instruments in the tropical Pacific to monitor stirrings of the thermocline.

The ocean is a mere film of salty water, some 4 kilometers deep, on a sphere whose diameter exceeds 12,000 kilometers. (If Earth were the size of an apple, the ocean would correspond to the peel of the apple.) In low latitudes the surface of the ocean has been bathing in intense sunlight for billions of years. Why, after all that time, is there so little warm water? So shallow is the surface layer of warm water, so much deeper the cold water below, that the average temperature of a column of water that reaches down to the ocean floor is barely above freezing point, even at the equator. Why is most of the ocean so cold? El Niño could become a permanent rather than an intermittent phenomenon if there were more warm water, so that the thermocline were deeper. Can global warming induce such a state of affairs and cause La Niña to disappear?

The ocean is heated inefficiently, from above, in contrast to the atmosphere, which is heated efficiently from below. Because the atmosphere is transparent to sunlight, solar radiation readily penetrates to the surfaces below the atmosphere, warms them up, and thus induces convection. That is the way we usually boil water—by putting a pot with water over, not under, a flame. The water at the bottom of the pot warms first, rises spontaneously—a process known as convection—and rapidly distributes heat throughout the pot. The ocean, by contrast, is heated from above because sunlight penetrates no more than a few tens of meters below the ocean surface. The downward transport of heat below the thermocline therefore depends on conduction, an extremely slow process in comparison with convection. Yet even the water in a pot under a flame (or broiler) ought to boil eventually, if we wait long enough. Likewise, the cold water in the deep ocean is in direct contact with the warm surface waters and, over millions of years, should have warmed up. Why has that not happened?

One of the first gentlemen to address this question was the versatile and peripatetic Benjamin Thompson from Rumford (now Concord), New Hampshire. Knighted by George III of England and named Count Rumford by a duke of Bavaria, this soldier, inventor, and gardener—he laid out Munich's famous *Englischer Garten*—was also a scientist. In an essay written in 1797, he proposed that the deep, cold water in low latitudes has its origin at the surface in high latitudes.[2] "If the water of the ocean, which, on being deprived of a great part of its Heat by cold winds, descends to the bottom of the sea . . . it will immediately begin to spread on the bottom of the sea, and to flow towards the equator, and this must necessarily produce a current at the surface in an opposite direction." Thompson realized that the ocean has a circulation that reaches from its surface down to its greatest depths. This ceaseless motion maintains the curious thermal structure of the ocean and profoundly influences the climate of our planet.

The picture of the oceanic circulation that has emerged from the integration of all the measurements gathered in different parts of the globe during the past two centuries indicates that the circulation has two main components, the one slow and deep, the other fast and shallow. The slow thermohaline circulation — sometimes referred to as a deep conveyor belt — involves the sinking of cold, salty water in high latitudes. After a millennium, approximately, the water has returned to the surface and is back at a region of sinking. The other main component of the circulation consists of several fast, shallow conveyor belts driven by the winds. Water parcels sink, in certain subduction zones in the subtropics, to a depth of a few hundred meters at most, and after a decade or two return to the ocean surface in equatorial and coastal zones where the winds induce upward motion. For a long time the oceanic circulation was regarded as steady, but over the past few decades interest has broadened to include the variability of the circulation, in order to understand phenomena such as El Niño and the likely effect of global warming on the oceans. This change in focus affected the manner in which oceanographers conduct their affairs, and it has brought the field to a difficult phase of transition.

In any science, measurements for the testing of theories and models are of vital importance. The oceanic measurements made up to the 1960s, mainly from single vessels on expeditions, were regarded as describing unchanging conditions and were often shown in atlases. Theoreticians turned to those books for descriptions of the phenomena they were trying to explain. The shift of interest to oceanic variability in the 1960s forced oceanographers to change their mode of operation; they started to organize international research programs involving several institutions. These programs had to be sustained for extended periods because the fluctuating and transient phenomena of interest, with the exception of the tides, never repeat themselves. For example, each El Niño is distinct, so we can learn about him only by observing him many times. Fortunately El Niño occurs every few years. Oceanographers therefore have had opportunities to study several different versions of El Niño over the course of a few decades—since the 1960s—and have been able to make rapid progress. (It is for the same reason that weather forecasting has advanced impressively—the weather changes often.) Oceanography is about to enter a new era because our interests are now turning to longer-term variability, changes in the properties of El Niño over several decades for example. This will require observations over many, many decades. But we need answers much sooner. That means that oceanographers, to test their theories, are obliged to take a keen interest in descriptions of climate fluctuations and climate changes in Earth's past. This will require close collaboration with the large and established community of paleoclimatologists. Oceanographers, who at first regarded themselves as explorers—the successors of da Gama, Magellan, and Drake—started participating in large, coordinated programs in the 1960s and now find themselves joining the even larger community of earth scientists. The ocean can no longer be viewed as a separate world but is a component of the "earth system," so integration of the community of earth scientists is essential. This is a major challenge because the different subdisciplines of that community have very different methods and cultures. This chapter is a summary of some of the developments in physical oceanography (which deals mainly with the motion of the ocean) up to now.

The Wind-Driven Circulation of the Upper Ocean

Initially, the surface currents of the ocean were of interest mainly because they affected the duration and safety of voyages. For example, in the eighteenth century mail packets took two weeks less to journey from New England to England than to make the reverse trip, because of the swift Gulf Stream. When Benjamin

Franklin learned about this from his cousin Timothy Folger, he improved the postal service — he was postmaster of the American colonies at the time — by producing a map of the Gulf Stream in 1770. Several decades later Matthew Maury, while serving as navigator on board the *Falmouth,* came to the conclusion that charts, not only of the Gulf Stream, but of oceanic conditions in general, could shorten sailing times significantly. At that time ship captains relied mostly on jealously guarded personal experience to navigate the high seas. Frequently they chose a direct course from port to port, without paying attention to contrary currents or doldrums. When Maury became superintendent of the Depot of Charts and Instruments of the U.S. Navy, in 1842, he decided to improve matters by transferring onto charts the vast amount of data on winds, currents, and sea conditions stored in ships' logbooks. He acquired additional data by offering ship captains copies of his charts, as they became available, in exchange for information they had to record on special forms Maury gave them. The charts — the first appeared in 1847 — proved a huge success. They enabled some vessels to reduce journeys from New York to California by as much as a month.

In 1855 Maury published *The Physical Geography of the Sea*, with chapters on navigation and winds. The book was immensely popular with the public, presumably because of the vivid descriptions of currents and other phenomena:

> There is a river in the ocean. In the severest droughts it never fails, and in the mightiest floods it never overflows. Its banks and its bottom are of cold water, while its current is of warm. The Gulf of Mexico is its fountain, and its mouth is in the Arctic Seas. It is the Gulf Stream.
>
> Matthew Maury, *The Physical Geography of the Sea*

Some scientists disapproved of Maury's fanciful theories and explanations that were often "unsustained by facts." Such was the opposition to Maury's activities that, shortly after his career as a marine scientist came to an end in 1861 — he returned to his home state, Virginia, when it seceded early in the Civil War — the newly organized National Academy of Sciences judged that his charts "embrace much which is unsound in philosophy and little that is practically useful." Publication of the charts ceased shortly afterward, but not for long, because the users of the charts judged the charts to be valuable, even though the academicians believed otherwise.[3] This incident is reminiscent of the attempt of professional scientists to halt weather forecasts in England in the 1860s on the grounds that the forecasts were without a sound scientific basis. The public seems to have a keener appreciation for the value of useful, practical information than do scientists.

The publication of Charles Darwin's *Origin of Species* in 1859 generated enormous public interest in biology, including marine biology. In response, a number of European scientific societies sponsored expeditions to explore conditions at and below the ocean surface, especially marine life at great depths. Obtaining the support of their governments was a relatively simple matter because, during the latter half of the nineteenth century, several European nations were keen on establishing worldwide empires and hence on having their ships sail the seas. Great Britain became the first nation to sponsor an expedition mainly for the purpose of gathering subsurface oceanic data when the HMS *Challenger* embarked on a four-year voyage in 1872. That expedition is usually regarded as having launched the science of oceanography. Expeditions by German, Norwegian, Russian, Italian, and French vessels followed shortly afterward. Although the goals of these cruises were primarily zoological studies, the measurements included vertical profiles of the ocean's temperature and salinity at numerous stations. That information enabled oceanographers to start investigating the oceanic circulation that Thompson had inferred in the eighteenth century. To do so they adopted methods very similar to those of weather forecasters, who plot contours of constant pressure to infer the winds.

Many people think of the ocean surface as being level, hence the term "sea level," but in reality, even on windless days when there are no waves, the ocean surface has slight undulations associated with its large-scale currents. For example, sea surface height is lower at Miami than at the nearby Bahama Islands because of the Gulf Stream, which flows northward between them. The ocean surface also slopes along the equator in the Pacific, with the water about half a meter higher in the west than in the east. Today, such information is readily available from satellites that measure variations in the height of the ocean surface with remarkable accuracy while circling the globe every ninety minutes. Long before the days of satellites, oceanographers knew of variations in the sea surface height because they used ingenious methods to map the contours of the ocean surface from measurements of temperature and salinity in the upper ocean. They exploited the fact that water expands when heated, so the higher the temperature of a column of water, the taller it is. (Salinity differences too affect the heights of different columns.) Hence it is possible, on the basis of temperature and salinity measurements, to draw maps that show lines of constant height at the ocean surface. Those are the equivalent of isobars on weather maps. Whereas isobars reflect atmospheric conditions aloft, the contours of sea surface height reflect conditions below the ocean surface.

Variations in sea surface height give rise to currents that at first flow from the peaks of the hills toward the valleys. Once in motion, the Coriolis force deflects the water until it flows essentially along lines of constant sea height (except at the equator, where the Coriolis force vanishes). Hence the contours of sea surface height are also the contours that the currents follow. The currents are mainly those of the upper ocean because the variations in surface height reflect mainly conditions in and above the thermocline. (Conditions below the thermocline are relatively uniform.)

Atmospheric pressure can vary considerably from one day to the next, so frequent measurements are necessary at the same location to determine the time-averaged value. Fortunately, subsurface temperatures and salinities in the ocean have less variability, so measurements gathered on numerous expeditions over many decades can be combined to create a global map of variations in the height of the ocean surface. (Whereas atmospheric conditions change continually, thus requiring an endless succession of weather maps, oceanic conditions were assumed to be so immutable that

measurements made at different times could all be used to produce one "frozen" map.) That was how oceanographers, by the
1960s, arrived at a reasonably accurate description of the major
currents in the upper layers of the oceans, in and above the thermocline. A schematic depiction of the surface currents is shown
in figure 14.1. Striking features of the oceanic circulation are the
asymmetrical gyres in each ocean basin: whereas the equatorward
flow is mostly in broad, slow, cold currents on the eastern side of
the basin, the return poleward flow is in warm, intense, narrow,
currents along the western side of the basin. The swiftness of the
rivers of warm water, the Gulf Stream and Kuroshio for example,
contrast sharply with their slow, cold counterparts, the California
Current and the Canary Current off northwestern Africa. How
are these currents related to the winds? There is one answer for
cold coastal currents — they are driven by local winds — and a
very different answer for the intense warm jets. The latter are in
response to the winds over the entire ocean basin.

In 1884 observations made from a ship, the *Fram*, frozen in the
ice of the Arctic Ocean, shed light on how the winds drive ocean
currents. On that occasion the Swedish oceanographer Vagn Walfrid Ekman noticed that the ship drifted, not in the direction of
the wind, but to the right of the wind. He attributed this curious
deflection to the earth's rotation — a consequence of the Coriolis
force. The wind exerts a force in one direction, but once in motion the Coriolis force deflects the ship and the ice toward the
right (in the Northern Hemisphere.) In the absence of ice, the
wind exerts a direct force on a very shallow, turbulent layer of
water that extends only a few tens of meters below the ocean
surface. In this Ekman layer, as it is known today, the wind force
plus the Coriolis force cause the water to drift to the right of the
wind. Sometimes this drift piles up the water in some regions.
The height of the ocean surface then increases in those regions
while it decreases elsewhere. That is the origin of the variations in
sea height that oceanographers infer from temperature and salinity measurements. Whereas the direct effect of the wind is confined to the shallow Ekman layer, the variations in sea height give

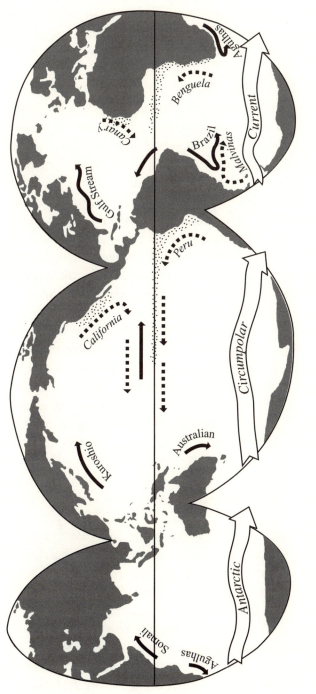

Figure 14.1. Some of the major wind-driven currents currents of the upper ocean. Those that flow poleward carry warm water; those that flow equatorward carry cold water. The Antarctic Circumpolar Current connects the three main ocean basins. The dots indicate regions of intense upwelling that are rich in nutrients and hence are rich fishing grounds.

rise to pressure forces that extend deeper, into the thermocline, so that the wind-driven currents of the upper ocean have a complex vertical structure and can be moving in different directions at different depths.

Deflections because of the Coriolis force have interesting consequences in coastal zones, those of California and Peru for example, where the winds are alongshore. In such regions the coastal currents are cold. At first the reason seemed obvious: the currents have their origin in high latitudes and flow toward the equator. The correct answer turns out to be more subtle. The currents do not simply flow in the direction of the alongshore winds but, because of the deflecting Coriolis force, also have a component away from the coast. This offshore flow causes the coastal upwelling of cold subsurface water rich in nutrients. That is why California and Peru have rich fishing grounds. The same happens in the vicinity of the equator, provided that the local winds are westward; in such longitudes the Coriolis force deflects the water northward in the Northern Hemisphere, southward in the Southern Hemisphere. This divergence of surface waters from the equator induces the upwelling of subsurface water. To sustain this upwelling, motion in the thermocline is toward the equator. To close this circulation, surface water sinks into the thermocline in subduction zones off California and northern Chile (and also off Northwest and Southwest Africa.) The surface waters in those latitudes are relatively warm and therefore do not descend to great depths, only a few hundred meters into the thermocline. The water in due course rises back to the surface in the upwelling zones.

This description of the shallow conveyor belt that effects an exchange of water between low and high latitudes emphasizes the vertical structure of the flow—poleward in the surface Ekman layer, equatorward below that in the thermocline. In a horizontal plane, the motion is in the asymmetrical gyres mentioned earlier. The salient features of those gyres, swift warm jets such as the Gulf Stream and Kuroshio, are puzzling because they seem unrelated to the local winds. In 1948 Henry Stommel, the dean of oceanographers during his long and highly productive career, took a significant step toward solving this riddle. He realized that, in a closed ocean basin, local winds cannot determine the currents everywhere. The circulation in such a basin must have features that satisfy global constraints. Suppose, for example, that the winds over a basin happen to drive local, southward currents everywhere. That happens to be the case in the subtropical Atlantic, the latitude of Miami for example. Why has southward flow across that latitude not emptied out the northern Atlantic? Because the Gulf Stream flows northward, making sure that the net flow across a circle of latitude is zero. What matters to that current are not the local winds over the current but the winds over the entire basin. Those winds, easterlies in low latitudes and westerlies in midlatitudes, exert a torque on the ocean, spinning the waters into a gyre. This is happening on a spherical planet that rotates rapidly from west to east, which is the reason for the pronounced asymmetry of the gyre, for the intense Gulf Stream in the west. A persistent torque will cause the gyre, or vortex, to spin faster and faster. What prevents that from happening? Stommel realized that, to satisfy this constraint of a balanced vorticity budget for the ocean, the Gulf Stream is again of central importance; it effects a dissipation of vorticity. Stommel's analyses, by means of a remarkably insightful, highly idealized model, made it clear why the Gulf Stream is a remarkable current.[4]

On the basis of measurements made on innumerable expeditions it was possible, by the middle of the twentieth century, to piece together how a water parcel in the upper ocean could travel from the northern Pacific, say, to the northern Atlantic. In the first leg of that journey the parcel sinks from the surface layers of the Pacific into the thermocline in a subduction zone off California. At first the parcel travels westward and in due course joins the Kuroshio, which carries it northward, then eastward. After rising back to the surface in a high latitude, the parcel proceeds southward, back to the subduction zone. After several such circuits, the parcel, while traveling westward along the southern branch of the gyre, veers toward lower latitudes. Eastward and westward it flows in the maze of tropical currents, countercurrents, and undercurrents until it reaches the equator, where wind-induced upwelling brings it to the surface. Soon it is drifting past exotic isles — Tarawa, Java, Sumatra — and into the Indian Ocean. There it may have to dawdle, waiting for the right monsoon winds to carry it toward Africa and into the warm Agulhas Current for the trek past Durban to the Cape of Good Hope.

There is an alternative route to the same destination. When at the equator in the Pacific, the parcel could drift into the Southern Hemisphere, all the way to the Antarctic Circumpolar Current, which joins the three major ocean basins, and which is driven by the roaring fourties and screaming fifties, the fierce westerlies around Antarctica. In that current the water parcel can circle the polar continent a few times. Upon emerging from the Drake Passage, it can escape by joining the northward Malvinas Current along the Argentinian coast. In due course it meets the warm, southward Brazilian Current near Buenos Aires, where the two embrace and tango eastward, toward Cape Town. (Satellite photographs actually show these two currents undulating in step as they flow eastward.)

In the cold waters off the western coast of South Africa, the parcel has an opportunity to join the gyre of the South Atlantic: northward in the Benguela, then westward to Brazil, at whose eastern tip there is a choice between returning to the south, to Buenos Aires, or proceeding to the north, across the equator, into the Gulf of Mexico. The Gulf Stream then becomes the first leg of a journey, in the Atlantic gyre, that includes subduction off northwestern Africa. After a few swings that cover the full width of the Atlantic, the water parcel escapes from the Gulf Stream and heads farther north toward Iceland.

During this odyssey the parcel travels around several gyres and simultaneously oscillates up and down, between the surface and the thermocline, to a depth of a few hundred meters at most. The wind-driven gyres therefore do not contribute to the maintenance of the very cold water at depth, below the thermocline. Only in the Antarctic Circumpolar Current, where the temperature difference between the surface water and the water at depth is slight, is there an obvious link between the upper and deep ocean. To explore other links we have to take into account that seawater is salty.

The Deep Thermohaline Circulation

Salt, which accounts for approximately 3 percent of seawater, increases the density of the ocean. (The Dead Sea is so salty, and hence so dense, that it is extremely easy to float on those waters.) If water is cooled until ice forms, then the salt is "squeezed out" of the water; an iceberg is therefore a source of fresh water. (Saudi Arabia will pay handsomely for the transportation of an iceberg to its shores.) When an iceberg forms, the water immediately below it becomes more saline and can become so dense that it readily sinks. Another means for increasing the salinity of water is through evaporation. When water evaporates from the ocean in a certain region and escapes into the atmosphere as the gas water vapor, then the salinity of the surface water in that region increases. Thus either evaporation or the formation of ice in high latitudes can create cold, salty surface water that readily sinks into the deep ocean. Regions where this happens are the sinking branches of the thermohaline circulation, which carries cold, salty water throughout the deep ocean.

To learn about the slow drift at depth, oceanographers assume that a particular water mass forms at the ocean surface in high latitudes, where the values of its temperature and salinity are set, then sinks to considerable depths and flows enormous distances from its source while retaining its characteristic temperature and salinity. The measurements reveal a surprising feature: in the Southern Hemisphere, very cold water sinks into the deep ocean all around Antarctica, but in the Northern Hemisphere, cold water sinks only in the far northern Atlantic, not in the northern Pacific or Indian Oceans. At their northernmost latitudes the Indian Ocean is too warm for the surface waters to sink, and the Pacific is too fresh. Only the Atlantic is sufficiently cold and saline for the surface water to sink. From the latter ocean the cold water spreads southward at depth, crosses the equator, and joins the Antarctic Circumpolar Current, which whirls around our globe in high southern latitudes. This eastward current, the only one that connects all three major ocean basins, enables the deep water to reach the Indian and Pacific Oceans.

A more detailed picture of that deep circulation became available after World War II because of new geochemical methods to track tracers in the deep ocean. Some of those tracers were by-products of the atomic bombs that we unwisely exploded in the atmosphere in the 1960s. Others included the man-made chemicals called CFCs, which are still destroying the ozone layer in the stratosphere. The distribution of these and other chemicals in the deep ocean corroborated estimates of the "age" of the deep water in different locations. (The "age" refers to the time since the water was last at the ocean surface.) The "youngest" deep water is that of the northern Atlantic. The "oldest" water, in the northern Pacific, has not been at the surface for close to a thousand years and as yet has no trace of CFCs.

Ultimately the deep water rises into the surface layers. Exactly where that happens is at present a major puzzle. This difficulty does not concern the cold water that appears at the surface in the oceanic upwelling zones of the subtropics and tropics — see figure 14.1 — and that comes from the thermocline at relatively shallow depths of a few hundred meters at most. Rather, we are ignorant of the manner in which water at depths of 1,000 meters and more returns to the upper ocean. For that to happen the water has to be heated, presumably by mixing the deep water with warmer water at shallower depths. But what is the source of energy for such mixing? The answer is still not known, but various possibilities are being investigated, including turbulence that is generated in the wake of motion over subsurface mountains on the ocean floor, the Mid-Atlantic Ridge for example. The matter is important because the rate at which waters of different temperatures are mixed in the deep ocean is one of the factors that determines the intensity of the thermohaline circulation and hence the extent to which that circulation influences climate.

One way in which the oceanic circulation affects climate is by redistributing heat. The ocean gains heat across its surface mostly in the cold upwelling zones of low latitudes, and it loses heat to the atmosphere in higher latitudes. The various conveyor belts, shallow and deep, transport heat from the regions of gain to the regions of loss. In the Pacific the transport is essentially symmetrical about the equator, poleward in both hemispheres, and is effected mainly by the shallow wind-driven conveyor belts. The Atlantic is very different because, unlike the Pacific, cold, saline water sinks in the far north. As a consequence both the shallow and deep conveyor belts contribute to the heat transport, which is from high southern to high northern latitudes, with a large increase in the neighborhood of the equator. This transport, especially the contribution made by the deep thermohaline circulation, is often cited as the reason for the mild winters of northwestern Europe. A recent study indicates that, in the absence of any oceanic heat transport, the climate of northwestern Europe would change very little.[5] Apparently the winds that bring relatively warm air to northwestern Europe are mainly a consequence of the global distribution of continents and mountains. Although the oceanic currents that transport heat northward are not important to the climate of northwestern Europe, the presence of a large pool of water, the northern Atlantic, is important. That pool has a high heat capacity, so in winter it releases to the atmosphere heat that was gained during the summer.

Even if the poleward transport of heat by the ocean has a small direct influence on climates in high latitudes, there are other ways in which that transport can be important. In the long term, the ocean must have a balanced heat budget; otherwise its temperature would rise steadily. Hence the loss in high latitudes must equal the gain in the tropics. If the loss were to cease for some reason — if there were to be a change in the heat budget — then more and more warm water would accumulate in the tropics, and the thermal structure of the ocean would be affected. The consequences could include a change in the depth of the equatorial thermocline and hence in the properties of El Niño. Such possibilities are under investigation.

The Response of the Oceans to Fluctuating Winds

The warm water that appears in the eastern equatorial Pacific during El Niño is associated with dramatic changes in the currents of the upper tropical Pacific Ocean. In the early 1960s essentially nothing was known about such variations in the oceanic circulation. Today we monitor those variations by means of an impressive array of instruments deployed symmetrically about the equator. (Figure 4.1 shows the TOGA array.) How did oceanographers come to the realization that a seemingly arbitrary line, the equator, in a remote part of the ocean, is of special interest?

The launching of *Sputnik* in 1957 and the subsequent intensification of the Cold War enhanced the military importance of the oceans and led to a huge increase in the resources available to oceanographers.[6] In the United States, for example, the navy supported the establishment and growth of oceanographic institutes at several universities, thus contributing to a significant increase in the number of scientists engaged in oceanographic research. Special funding for the International Decade of Ocean Exploration — the 1970s — facilitated joint projects between different nations. Several of those projects explored the variability of oceanic conditions. For this purpose, expeditions by single vessels are of very limited value. To measure the fluctuations of currents and temperatures continuously, over large regions for extended periods, requires coordination of the efforts of investigators at different institutions, often in different countries. The oceans are so enormous that, initially, only limited areas could be studied for limited periods. Sometimes the place and time were determined in an opportunistic rather than a scientific manner.

In the early 1970s meteorologists proposed a huge field program to study atmospheric conditions over the tropical Atlantic during the summer of 1975. The program involved the deployment of an armada of ships (to serve mainly as platforms for atmospheric measurements of tropical clouds and related phenomena), so oceanographers were invited to participate. At the time, the available information concerning oceanic variability was still so scant that many oceanographers doubted whether measurements for such a short period would reveal anything interesting. They argued that only measurements over a period of several years would tell us anything about oceanic variability and therefore declined to participate. A few oceanographers nonetheless joined the field program and serendipitously made an exciting discovery: the equatorial currents were found to meander energetically with a period of a few weeks.[7] At about the same time Richard Legeckis detected similar undulations in satellite photographs of sea surface temperature patterns in the Pacific.[8] A simple theory for these meanders — they arise spontaneously when the equatorial currents become too intense — yielded results in accord with the measurements.[9]

Shortly afterward Carl Wunsch and Adrian Gill demonstrated that known theoretical results concerning curious waves confined to the vicinity of the equator can explain puzzling oscillations in sea level as measured by tide gauges at various islands in the tropical Pacific.[10] (The meteorologist Taroh Matsuno first developed a theory for these waves; Dennis Moore first explored them in an oceanic context.)[11] Although these sea level fluctuations, and the meanders of equatorial currents, are phenomena of secondary importance, their discovery generated considerable excitement for two reasons. The unexpected phenomena established that even measurements over periods as short as a few days and weeks can document energetic fluctuations in the equatorial Atlantic and Pacific. Furthermore, the agreement between the measurements and theories provided welcome reassurance that theories concerning oceanic variability are not entirely "dream-like" but are relevant to observable phenomena.

These findings motivated oceanographers to design a number of programs, each involving close collaboration between observers and theorists, to explore all three tropical oceans. Exciting discoveries concerning waves and currents with curious properties came at a rapid pace. There was a need for quick communication between participants in different projects because the phenomena, although they may seem unrelated, are all governed by the same physical principles. Effective, rapid communication was achieved in two ways. David Halpern skillfully edited a very successful newsletter in which scientists presented short summaries of their results in advance of publishing the results in formal journals. (The World Wide Web and e-mail did not exist yet.) Convivial Dennis Moore chaired a series of informal scientific meetings that were very effective forums for the discussion of new results, observational and theoretical, concerning phenomena not only in the Indian Ocean but in the Atlantic and Pacific too. The participants were from oceanographic centers across the globe. Philippe Hisard, Christian Colin, and their associates from Nouméa and Papeete reported on the perplexing fluctuations of equatorial currents in the far western equatorial Pacific, where the monsoonal winds reverse direction seasonally. Jacques Merle and his colleagues in Dakar and Abidjan described the Atlantic counterpart of El Niño, the occasional warming of the otherwise cold western coasts of tropical Africa. David Anderson and Adrian Gill in Cambridge (England), Sulo Gadgil in Bangalore, and Toshio Yamagata in Tokyo developed theories that attempt to explain some of these phenomena.

Theoretical interest in the variability of tropical currents stems from a question George Veronis and Henry Stommel asked in the 1950s: if the winds over the ocean were to stop blowing suddenly, how long would it be before the Gulf Stream disappeared? (Alternately, if the winds abruptly start blowing over an ocean that previously had been motionless, how long before a Gulf Stream appears?) The winds at first drive local currents everywhere, as if the ocean were unbounded. In due course, constraints imposed by the presence of coasts and the boundedness of the ocean basins influence the oceanic response to the winds. At that stage a Gulf Stream is likely to appear because it satisfies global constraints as explained earlier. Veronis and Stommel calculated that the generation of the Gulf Stream along the western boundary of the basin starts after the arrival of waves excited along the eastern boundary of the basin. These Rossby waves, as they are known, propagate westward along the thermocline and effect the oceanic adjustment to a change in the winds. They travel so slowly in midlatitudes that, for a basin the size of the

Atlantic, it takes on the order of a decade to generate a Gulf Stream from a state of rest.[12] The speed of these waves increases with decreasing latitude and, in the simplest theories, is infinitely fast at the equator. This intriguing result suggested to James Lighthill that the Somali Current along the eastern coast of equatorial Africa is a speeded-up version of the Gulf Stream, capable of reversing direction seasonally because it is in low latitudes, where the oceanic adjustment to changing winds is very rapid.[13] Would it be possible for oceanographers to observe, over a period of months near the equator, processes that occur over decades in high latitudes? Could the appearance of warm water in the eastern equatorial Pacific during El Niño be another example of the rapid adjustment of the equatorial oceans to changes in the winds? Confirmation of this last conjecture came from an unexpected source, tide-gauge records.

Tidal records serve mainly local, commercial purposes, but a large number of such records from islands and coastal ports across an ocean basin also provide information about changes in large-scale oceanic conditions. This is so because sea level fluctuates in response to tidal forces and also in response to temperature changes. If the depth of the thermocline increases, the average temperature of a column of water increases; the local sea level will then rise because water expands when it warms up. In the 1960s Klaus Wyrtki of the University of Hawaii analyzed tide-gauge records from ports along the American and Asian coasts, and on islands scattered across the tropical Pacific, and made a remarkable discovery: the warming of the eastern tropical Pacific during El Niño is accompanied by a rise in sea level in the east and a simultaneous fall in sea level in the west.[14] Wyrtki interpreted this to mean that the thermocline deepens in the eastern Pacific during El Niño while it shoals in the west, and he concluded that the warm water lost in the west flows eastward to cause the rise in sea level there.

With the exception of the Galápagos Islands, which are 1,000 kilometers off the coast of Ecuador, the eastern half of the equatorial Pacific has almost no islands that can serve as platforms for tide gauges. In the early 1970s the oceanic signature of El Niño in that vast region was a mystery. Presumably the east-west redistribution of the warm waters of the upper ocean during El Niño involves dramatic changes in the oceanic currents, but nothing was known about those changes. To get information it would be necessary to deploy instrumented moorings that could remain unattended for prolonged periods. One approach would be to suspend instruments from taut wires, between an anchor on the ocean floor and a balloon at the surface. The design of such moorings is a major engineering challenge. The taut wire is under enormous stress because intense equatorial currents flow in opposite directions at different depths. Furthermore, the wire gets stretched relentlessly when wind waves cause the ocean surface to rise and fall. The first moorings were failures; they broke loose from their foundations on the ocean floor and drifted away. After a while, persistent David Halpern (now at the Jet Propulsion Laboratory in Pasadena, California) succeeded. The instruments he had in place throughout the exceptional El Niño of 1982 and 1983 not only documented the passage of various equatorial waves that theoreticians had proposed but also provided a detailed description of remarkable changes in the equatorial currents, including the temporary disappearance of the mighty Equatorial Undercurrent.[15]

In the mathematical analyses of Veronis and Stommel, the oceanic adjustment to changes in the winds is effected by Rossby waves that are infinitely fast at the equator. Matsuno and Moore's theoretical investigation of this singular behavior revealed the possibility of a family of very rapid waves that are trapped about the equator. Some are modified versions of the westward Rossby waves; a very important additional wave, possible only along the equator, where it travels eastward, and along coasts, is the Kelvin wave. To gain an understanding of exactly how these waves effect the oceanic adjustment, theoreticians use idealized models to investigate idealized situations. The goal of such studies is not realistic simulation of what actually happens but the building of a "vocabulary" of certain basic concepts that can be applied to a variety of situations. Consider, for example, a very simple model in which a closed ocean basin has an upper warm layer of constant temperature floating on an infinitely deep cold layer. Initially there are no winds, so the ocean is at rest and the thermocline is horizontal. Suddenly spatially uniform westward winds turn on and then remain steady. After a certain time the winds maintain a sloping thermocline, downward to the west, but drive no currents at all. This is a most unrealistic state of affairs — in reality winds drive currents — but this particular model is of enormous value for its description of the journey, not the final destination. From that description we learn that, even though the winds blow uniformly over the entire basin, most of the action is along the equator. Fascinating events immediately after the abrupt onset of the winds include the appearance of an accelerating equatorial jet and waves that travel swiftly, eastward and westward, along the equator. Those waves first arrest the acceleration of the jet, then reverse and finally eliminate the jet, leaving behind a sloping thermocline. While this is happening, signals spread poleward along the eastern boundary of the ocean basin (the western coast of the Americas if we are dealing with the Pacific). Calculations in which the winds are confined to the western side of the ocean basin show that a jet appears only where the winds are blowing, and that equatorial Kelvin waves penetrate into the eastern side of the basin, to the coast of the Americas, where signals propagate poleward.

These results clarify how, during the early stages of El Niño, a relaxation of winds near the date line can affect the thermocline and hence sea surface temperatures in the eastern Pacific by means of waves that travel along a guide, the equator. The special significance of that line — and the rationale for an array of instruments along the equator — emerged from calculations with highly idealized models. Yoshida, McCreary, O'Brien, and their students pioneered these studies.[16] Cane and Sarachik, in a comprehensive series of papers, developed general methods for calculating the response of that idealized ocean, consisting of two layers of fluid, to any winds.[17] The model discussed thus far has a uniform temperature for its upper layer and hence has no sea surface temperature patterns. This deficiency, and the need to investigate the transient Equatorial Undercurrent of the Indian Ocean — how long after the sudden onset of winds before that current appears? — called for more sophisticated ocean models. This need was underlined when observations revealed that the most prominent currents along the equator are always eastward, independent of the direction of the wind. Eastward winds in the central Indian Ocean generate an intense eastward surface jet; when the westward winds over the Pacific and Atlantic are strongest, so is the eastward Equatorial Undercurrent, which can even drag the surface flow eastward. This is known as a nonlinear response to the winds.

To cope with the complex vertical structure of equatorial currents and undercurrents requires a model with not one or two but a large number of layers and hence requires the resources of supercomputers. In the 1960s, at the Geophysical Fluid Dynamics Laboratory in Princeton, where meteorologists were developing computer models for the simulation of weather and climate, Kirk Bryan and Michael Cox started building the oceanic counterpart of those models.[18] Stommel had already demonstrated how, in an idealized, closed ocean basin confined to one hemisphere, the oceanic circulation depends on local winds and also on global constraints. Bryan and Cox started developing a tool to investigate the same problem for the global ocean. Given the very limited computer resources at the time, this was a very ambitious project. At first some oceanographers were skeptical of this tool, but today the Oceanic General Circulation Model is used widely to investigate a broad range of problems. A successful simulation of the growth and decay of El Niño of 1982 contributed to the general acceptance of the model. When forced with the observed winds, the model realistically reproduced the observed changes in surface and subsurface temperatures and in the various currents, including the temporary disappearance of the otherwise swift Equatorial Undercurrent.[19]

Oceanographers made such rapid progress that they were able to document, explain, and simulate all the oceanic aspects of the evolution of El Niño in 1982 and 1983. Although they failed to alert the public to the developing El Niño of 1982, they quickly took the necessary steps to do so in future when Stan Hayes (of

NOAA in Seattle) deployed the TOGA array of instruments in the Pacific.[20] To cope with the rapidly increasing volume of data from the Pacific, Ants Leetmaa (then at NOAA in Miami) started using the Oceanic General Circulation Model for operational purposes.[21] Thus did oceanography acquire its first operational activity, the counterpart of what meteorologists have always had. (Discussion of the next step, coupling the oceanic and atmospheric models to predict climate fluctuations such as El Niño, is deferred to the next chapter.)

The General Circulation of the World Ocean

Over the past few decades our understanding of El Niño progressed rapidly because that phenomenon occurs relatively often. This has two important implications: several events can be observed over the span of a few decades, and the duration of an event is so short, on the order of a year, that the basic thermal structure of the ocean remains essentially constant over that time and can be regarded as a given. El Niño can therefore be studied without paying much attention to the complex processes that maintain the oceanic thermal structure — the thermocline for example — on timescales of decades and longer. Growing interest in long-term (decadal) changes in the properties of El Niño, and in the possible impact of global warming on that phenomenon, is prompting oceanographers to develop theories and models for the processes that maintain the oceanic thermal structure. This means that the models have to take into account both the deep and the shallow conveyor belts of the oceanic circulation. At present the connections between these two components are poorly understood. Tools to deal with this issue are being developed — they are refined versions of the computer models that Kirk Bryan and his collaborators originally developed at Princeton. How do we obtain data sets that provide stringent tests for the models? Realistic simulation of present conditions is inadequate because the models are needed to reproduce different worlds — one in which global warming has taken effect, for example. At present oceanographers are conducting several field programs to address issues such as the connection between the deep and shallow components of the global oceanic circulation, but it will take many decades if not a century to acquire data sets that describe decadal variability. Fortunately, alternative data sets are available — those that describe climate changes in the earth's past, the topic of chapter 16.

FIFTEEN

Reconciling Divergent Perspectives on El Niño

Glendower:
I can call the spirits from the vasty deep.
    Hotspur:
Why, so can I, or so can any man; But will they come when you do call for them?
    —William Shakespeare, *Henry IV*, part I

Asking the right question is the key to solving a difficult scientific problem. This is no trivial matter, because any question reflects implicit assumptions. The right question is based on the correct hypotheses, and it often emerges only after several others have been considered. In studies of El Niño, scientists at first asked

1. What causes El Niño?

After a while some turned their attention to a different question:

2. What factors determine the properties (period, amplitude, etc.) of the oscillation between El Niño and La Niña?

These questions are based on very different assumptions concerning the character of El Niño, assumptions that reflect different perspectives on the nature of time. Those who address question 1 regard El Niño as a departure from "normal" conditions. The phenomenon starts to develop at a certain time because of "triggers" that lead to the growth and subsequently to the decay of "anomalous" atmospheric and oceanic conditions. Each El Niño has a beginning and an end, so time is perceived as an arrow that moves forward, in a definite direction. Question 2 implies an entirely different perspective: El Niño and La Niña are regarded as the complementary phases of a continual oscillation with no beginning or end. We are dealing with one of time's unending cycles, so inquiries about "triggers" that initiate developments are pointless. Of prime interest are the factors that determine the properties of the oscillation, such as its period.

Those who address question 1 and those who address question 2 have such different perspectives on El Niño that they give very different explanations for the irregularity of the Southern Oscillation (and hence for the uniqueness of each El Niño). They also disagree about the extent to which El Niño is predictable. Some believe that he will readily come when our prophets call him from "the vasty deep." Others are far less certain. To trace the origin of these divergent perspectives, we have to review the recent history of research on El Niño.[1] It then becomes clear that the disagreements can be resolved by means of a compromise, by accepting that time can be viewed as both an arrow and a cycle.

The data concerning El Niño were so scant in the 1960s that it seemed reasonable to regard the phenomenon as an occasional departure from "normal" conditions. In that case "1" is the appropriate question to address. Presumably El Niño appears when the ocean-atmosphere interactions proposed by Bjerknes amplify an appropriate initial disturbance or "trigger."[2] Klaus Wyrtki at the University of Hawaii analyzed the available data and identified the "trigger" to be an intensification of the trade winds over the tropical Pacific, followed by the collapse of those winds.[3] This result motivated theorists to study the oceanic response to an abrupt change in the winds, in the belief that such a change leads to the development of El Niño. So prevalent was this view of El Niño as the response to a "trigger" that the strategies some scientists proposed for detailed observations of El Niño were very similar to those that are used to observe sporadic phenomena such as hurricanes. In the 1970s there were plans to keep instruments in reserve in certain laboratories and to be ready to rush to the tropical Pacific should El Niño start developing. In this spirit, the tentative identification of certain precursors of El Niño in 1975 led to the rapid organization of two cruises to the equatorial Pacific. The alert was premature.[4] El Niño failed to appear that year but did develop in 1976. A few years later, in 1982, the opposite happened. The absence of any precursors lulled everyone into believing that no El Niño was imminent. The most intense El Niño in more than a hundred years nonetheless developed in 1982.

In the 1980s the availability of new data sets revealed the occasional presence of westerly wind bursts along the equator for periods as long as a month. The idea that these bursts could be the "triggers" of El Niño gained ground when mathematical studies clarified why the ocean-atmosphere interactions proposed by Bjerknes are confined to the neighborhood of the equator. Of central importance is the absence of the Coriolis force at that special latitude. Along that line, a westward wind drives the warm surface water westward and creates an east-west temperature gradient that reinforces the wind.[5] Such interactions between the ocean and atmosphere are impossible far from the equator because the Coriolis force deflects the oceanic motion from the direction in which the wind is blowing; if sea surface temperature gradients are created, they do not reinforce the winds. The equator is where ocean-atmosphere interactions can amplify certain "triggers" or precursors of El Niño, the westerly wind bursts near the date line for example. Such bursts were indeed of central importance to the initiation of the exceptionally intense El Niño episodes of 1997.[6] However, in the records, "triggers" similar to those of 1997 fail to produce El Niño on many occasions. Furthermore, El Niño sometimes appears even in the absence of wind bursts. Why do some wind bursts result in El Niño while others are ineffective? The same question can be asked of oceanic Kelvin waves that travel eastward along the equator. Some people are under the impression that those waves are essential to the development of El Niño. In reality, Kelvin waves are observed far more often than are El Niño episodes, and sometimes those episodes develop in the absence of explicit Kelvin waves. Apparently we need to continue the search for the cause of El Niño. Or maybe we need to ask different questions about El Niño. Question 1 is not proving fruitful.

By the early 1980s we had far more information about El Niño than we did in the 1960s, not only because of the additional measurements that were made during the intervening years, but also because of the recovery of earlier measurements made by commercial vessels over many decades. The concerted efforts to retrieve those measurements from old ships' logs and files proved immensely valuable. They provided a new perspective on El Niño. To some scientists the data suggested that El Niño is not a sporadic occurrence — an occasional departure from "normal" conditions — but is part of a continual oscillation, the Southern Oscillation.

Some periodic phenomena, the tides and the seasonal cycle for example, have definite causes. Some have no apparent cause; they seem to appear spontaneously. A familiar example is a pendulum that swings back and forth. We take this oscillatory motion for granted, and rather than ask why the pendulum swings, we investigate the factors that determine its period and amplitude. A very slight breeze when a door is opened could be the disturbance that initiates the oscillations, but we take far less interest in that disturbance than in the period and other properties of the oscillation. We similarly take it for granted that, by simply dropping a pebble into a pond, we can generate a most remarkable phenomenon: perfectly concentric circles that remain perfectly concentric while they radiate outward. The properties of these waves are of far greater interest than the detailed shape of the pebble. In the 1980s some scientists proposed that the oscillation between El Niño and La Niña is as spontaneous as the swings of a pendulum, that the phenomenon is a natural mode of oscillation of the interacting ocean and atmosphere. They therefore turned their attention from the possible disturbances that initiate the oscillation to the properties of the oscillation. What are the factors that determine its period and amplitude? They turned to questions that had puzzled Jacob Bjerknes in the 1960s. Bjerknes had realized that interactions between the ocean and atmosphere could amplify a modest initial weakening of the trade winds into El Niño, but he could not explain why those conditions, once established, did not persist indefinitely.

Laboratory experiments contribute enormously to our understanding of many phenomena. For example, experiments with a pendulum establish that the length of the string, not the weight at the end of the string, determines the period. In the case of the Southern Oscillation, experiments are ruled out, but scientists can turn to mathematical models that simulate interactions between the ocean and atmosphere and can use those models to perform "numerical experiments." The first step is to develop a model capable of reproducing an oscillation that resembles the Southern Oscillation. The effect of different factors on the properties of the simulated oscillation can then be determined, so that it becomes possible to see the observed Southern Oscillation in the broader context of many possibilities. By the 1980s scientists had developed the two components of such a model: oceanic models that reproduce the oceanic response to changes in the winds, and atmospheric models capable of simulating how the winds are affected by changes in sea surface temperatures. If coupled together, would these components spontaneously reproduce oscillations between El Niño and La Niña? Hotspur phrased the question differently: "Will they come when you do call for them?"

The message that summons the spirits should be as simple as possible, without distracting asides. That was the philosophy adopted by Mark Cane and Steve Zebiak. In the early 1980s, at the Massachusetts Institute of Technology, they constructed an elegant and relatively simple coupled model to simulate interactions between the atmosphere and ocean.[7] Given sea surface temperature patterns in low latitudes, their simulated atmosphere reproduces that which matters most to El Niño, the tropical winds, without introducing extraneous features such as the Jet Streams, cyclones and anticyclones, warm and cold fronts. Their simulated ocean is similarly focused. It regards the thermocline as a given, thus excluding the circulation that maintains that feature. Instead the model concentrates on the rise and fall of the thermocline in response to changing winds and on the associated surface temperature patterns. When Cane and Zebiak coupled the idealized ocean and atmosphere—the winds inducing surface temperatures

that in turn affected the winds—spontaneous oscillations between El Niño and La Niña appeared! The "spirits from the vasty
deep" came when they were called.

The oscillation that the idealized model simulates resembles the
observed Southern Oscillation in many respects. For example, the
period is four years approximately. But which factors determine
that period? The model, though highly idealized, is so complex
that an answer was not immediately evident, not until Paul
Schopf and Max Suarez of NASA (and then shortly afterward
and independently David Battisti and Tony Hirst of the University of Washington) identified the "memory" of the ocean as the
key factor and introduced the idea of a delayed oscillator.[8] (As
explained in chapter 1, we experience this type of oscillation
when taking a shower in a bathroom with old-fashioned plumbing.) Of crucial importance is the difference between the rapid
atmospheric response to changes in surface temperature patterns
and the slow oceanic adjustment to a change in the winds. The
state of the atmosphere at any time depends on the surface temperatures at that time, but the state of the ocean at any time
depends, not only on the winds at that time, but also on earlier
winds. This "memory" of the ocean is in the form of undulations
of the thermocline, undulations that are most rapid and coherent
near the equator. It became clear that, to anticipate future developments, subsurface oceanic conditions have to be monitored, especially in the neighborhood of the equator.

In the highly idealized models, one El Niño is identical to the
next, La Niña is the mirror image of El Niño, and the Southern
Oscillation is perfectly regular and hence perfectly predictable. In
such a timeless world the oscillation has no irregularities. Why, in
reality, are there irregularities? Why is each El Niño distinct? For
answers we need to perform "numerical experiments" with a
model that simulates the Southern Oscillation.

We choose as our "apparatus"—our computer model of the
coupled ocean-atmosphere[9]—one that has no weather, that is free
of random atmospheric disturbances. An initial disturbance is
therefore necessary to initiate interactions between the two media. To that end a burst of westerly winds is imposed near the

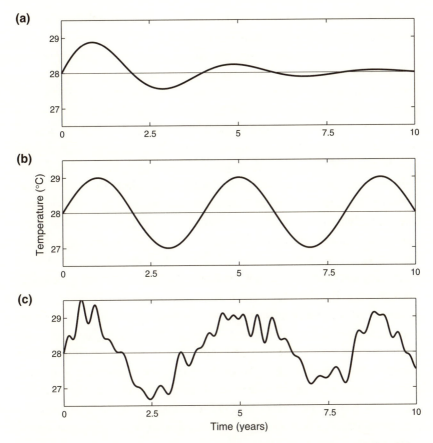

Figure 15.1. Changes in sea surface temperature along the equator (in degrees centigrade) in a coupled ocean-atmosphere model similar to that of Neelin.[9] The strength of the coupling between the ocean and atmosphere increases from the top to the middle and bottom panels.

date line during the first month of each experiment. Subsequent developments are very different in the three experiments shown in figure 15.1. Each has a Southern Oscillation, but it is damped in the first experiment, self-sustaining in the second, and unstable in the third. This is a consequence of increasing the intensity of ocean-atmosphere interactions from one experiment to the next; the effect of given winds on sea surface temperatures increases from the first to the third experiment. (To induce such changes, the depth of the specified thermocline in the model can be increased from one experiment to the next.)

In the first experiment (top panel of figure 15.1) ocean-atmosphere interactions are very weak. A modest El Niño develops but soon attenuates, whereafter "normal" conditions return. For further developments we will have to wait for a new burst of winds. In this case the Southern Oscillation is damped. It is as if we were dealing with a pendulum that is barely capable of swinging, because it is immersed in water rather than air.

In the second experiment (figure 15.1, middle) the interactions between the ocean and atmosphere are more intense. We now find ourselves in a world very different from that of the first experiment. As before, a brief, imposed wind burst initiates developments, but soon there is no evidence of a beginning, only of a continual cycle that endlessly repeats itself in a perfectly regular, perfectly predictable oscillation.

Ocean-atmosphere interactions are most intense in the third experiment, shown in figure 15.1 (bottom panel). Now the oscillations are so unstable that they become irregular. This chaos persists even in the absence of any further externally imposed disturbances, a state of affairs very different from that in the figure's top panel. For that first experiment to have a continual, irregular oscillation, random external disturbances — bursts of westerly winds, for example — have to be introduced repeatedly.

These experiments indicate that several mechanisms can cause the irregularities of the Southern Oscillation. A highly unstable oscillation has its own internal mechanisms that generate chaos and irregularities, as in the bottom panel of figure 15.1. A strongly damped oscillation, on the other hand, requires an external source of noise to be irregular, as in the top panel. These results shed light on the divergent perspectives on El Niño mentioned earlier. Those who address question 1, who believe that consecutive El Niño episodes are unrelated, that each is a departure from "normal" conditions, are inclined to favor the top panel in figure 15.1 as the one most relevant to reality. Those who address question 2, who regard El Niño as part of a continual oscillation, favor the other two panels. This matter is of far more than academic interest because it has important implications for the predictability of El Niño. In the world of the bottom panel, predicting El Niño is very similar to predicting the weather because forecasts are limited by errors in the description of the initial state. So unstable is the system that those errors rapidly amplify, thus invalidating the forecast after a short time. In the world of the top panel, predictability is also limited, but for a very different reason: the time of occurrence of a random wind burst that initiates developments cannot be anticipated. Furthermore, once El Niño has started to develop, further unanticipated disturbances are possible. At best, we can know the statistics of the random disturbances so that probabilistic forecasts are possible. To develop a model for the prediction of El Niño, it is therefore important to know which of the panels in figure 15.1 is most relevant to reality.

A number of investigators repeated the series of experiments in figure 15.1 after making a few changes, one of which is particularly important: they provided the atmospheric component of the coupled ocean-atmosphere model with "weather" by adding to the simulated winds the atmospheric noise observed in the tropical Pacific.[10] (To determine the noise, the component of the observed wind field that is correlated with sea surface temperature changes is subtracted from the total observed winds; the remainder is the noise.) In each of the counterparts of the panels of figure 15.1, an irregular interannual oscillation now appears. Which simulated interannual oscillation has statistics—the position, width, and intensity of a peak in an energy spectrum for example—comparable to the statistics of the observed Southern Oscillation? These calculations indicate that the most realistic oscillation in this statistical sense is weakly damped and is sustained by atmospheric noise.

There is persuasive evidence that the Southern Oscillation is neither strongly damped nor highly unstable. If it were strongly damped, then El Niño would appear only in response to an appropriate disturbance such as a westerly wind burst. It would then be difficult to explain why similar wind bursts initiate El Niño on some occasions but not others. If it were highly unstable, westerly wind bursts would not matter. It appears that the

Southern Oscillation is weakly damped and is sustained by random disturbances. The effect of a wind burst therefore depends on its timing — on when, during a cycle, the burst appears. Imagine a pendulum swinging back and forth, struck by modest blows at random times. A blow at the right time amplifies the swing; at the wrong time it dampens the swing. If this is an analogy for the observed Southern Oscillation, then we can explain why similar wind bursts induce different responses at different times, and we can also explain why the Southern Oscillation has a distinctive timescale.

These results imply that the disagreements between those who address question 1 and those who address question 2 can be resolved by means of a compromise, by accepting that time can be viewed as both an arrow and a cycle. The Southern Oscillation involves both a long and a short timescale: the period of the oscillation is several years, but major developments can occur far more rapidly, over a period of weeks or months. The Southern Oscillation has some long-term predictability because the presence of a cycle enables us to anticipate years in advance when El Niño is likely to occur. Between 1982 and 2002 the cycle was particularly regular, and El Niño appeared every five years. Its amplitude, however, varied enormously during that period. El Niño was exceptionally intense in 1982 but weak in 1992. Such differences are attributable to random wind bursts, so the intensity of El Niño cannot be predicted deterministically. Given the statistics of the wind bursts, scientists can predict the probability that El Niño will have a certain intensity. In 1982 and again in 1997 a succession of wind bursts occurred as the Southern Oscillation was about to enter its warm phase. An accurate forecast for El Niño of 1997 would have assigned a low probability to the event that actually developed because the succession of wind bursts early that year is an infrequent occurrence.[11] Once that strong episode was well under way, developments over the subsequent several months could be anticipated with reasonable confidence. That was why, in June 1997, scientists could alert Californians to the possibility of heavy rains in early 1998.

To compare the coupled ocean-atmosphere to a simple pendulum is a gross oversimplification because the pendulum has only one mode of oscillation—the swings, back and forth, of the heavy weight at the end of a string. The ocean and atmosphere coupled together are far more complex and are capable of many modes. The model of Cane and Zebiak is a valuable tool for exploring the different possibilities because that model regards the thermocline as a given. This means that the depth of the thermocline can be specified to be as observed in the Pacific today, or to be much shallower or much deeper (as would be the case in a world colder or warmer than ours is at present). Do different worlds favor different types of oscillation? David Neelin and his colleagues at the University of California in Los Angeles addressed this question and demonstrated that many types of oscillation are possible. Of special interest is a mode that the seasonal variations in the intensity of sunlight excite in the eastern equatorial Pacific. Even though the sun "crosses" the equator twice a year, sea surface temperatures near the Galápagos Islands are at a maximum once a year, in April; six months later temperatures are at a minimum. The region of warmest surface water, onto which the winds converge—westerly winds to the west, easterly winds to the east—drifts westward along the equator and in due course is followed by a patch of cold water that drifts westward. This mode, unlike the delayed oscillator, involves neither vertical movements of the thermocline nor any "memory" of the ocean.[12] Of central importance is a local interplay, between the winds and surface temperatures, that exploits the difference between east and west: at the equator easterly winds drive divergent currents that decrease surface temperatures; westerly winds do the opposite. We refer to this as the local mode.

A shallow thermocline favors the local mode, which tends to have a relatively short period; a deep thermocline favors the delayed oscillator with a longer period. If the depth of the thermocline were to increase gradually, then there would be a smooth transition from one type of oscillation to the other. There is evidence that such an increase in the equatorial Pacific, from the 1960s and 1970s to the 1980s and 1990s, was associated with a decrease in the frequency of occurrence of El Niño, from every three years to every five years approximately. This means that the observed Southern Oscillation corresponds, not to a strict delayed oscillator, but to a hybrid mode with properties that are changing gradually on a timescale of decades.[13] Information about such changes is now becoming available from sediments in lakes and from corals that lived for a few decades, long ago, and that have annual growth rings similar to those of trees. Those records of El Niño in the distant past reveal that at one time, some 7,000 years ago, El Niño disappeared temporarily. Models of the type that yield the results in figure 15.1 indicate that such a disappearance is possible when the thermocline is so deep that the winds are unable to bring cold water to the surface. Ocean-atmosphere interactions are inhibited under such conditions. Information about the depth of the thermocline some 7,000 years ago will provide a valuable check on the validity of the model results.

Although several of the main features of El Niño have been explained persuasively, some major puzzles remain unsolved. For example, it is not known why El Niño is often brief and is then followed by prolonged La Niña conditions. In 1982 and again in 1997, El Niño lasted for about 1 year, whereafter La Niña persisted for about 4 years. On such occasions, once El Niño started to emerge, subsequent developments and the transition to La Niña were relatively easy to anticipate. By contrast, the transition from La Niña to El Niño is proving much more problematic. In many models La Niña is the inverse (or negative) of El Niño, but that is clearly not the case in reality. The two have intrinsic differences that are yet to be explored. The sea surface temperature patterns that characterize El Niño and La Niña provide clues. Whereas temperatures tend to be uniformly warm across the tropical Pacific during El Niño, they have striking asymmetries relative to the equator during La Niña. The ocean-atmosphere interactions that give rise to the north-south asymmetries in the eastern tropical Pacific have thus far received far less attention than have the interactions involving the east-west redistribution of warm water. The former depend on a shallow thermocline in the east and therefore are more prominent during La Niña than El Niño. Could it be that north-south interactions, once they intensify during La Niña, tend to favor persistence of those conditions? What are the other factors that can cause La Niña to be more prolonged than El Niño?

Models that specify the depth of the thermocline are powerful tools for exploring how El Niño is affected when the depth of the thermocline changes. However, those models do not tell us why the depth of the thermocline changes or why, in the world of today, the thermocline is so shallow. This means that the models cannot be used to study a type of El Niño very different from the one that appears every few years. The one with which we are familiar involves a horizontal redistribution of the warm surface waters of the upper tropical Pacific Ocean. An eastward movement of the warm waters along the equator causes temperatures in the eastern equatorial Pacific to rise. Such an increase in temperatures can also occur in a very different manner, when the thermocline deepens everywhere. The latter mechanism, which requires an increase in the flux of heat from the atmosphere into the ocean, is of interest because it may be involved in the decadal modulation of El Niño and could be an aspect of global warming induced by the rise in the atmospheric concentration of greenhouse gases. The next chapter (16) discusses conditions on Earth earlier than 3 million years ago, when El Niño was a perennial rather than intermittent phenomenon, possibly because of a deep thermocline.

Thus far, most theoretical studies of El Niño have used relatively simple models that cleverly bypass difficult issues such as the processes that maintain the oceanic thermocline. That feature is simply specified, and the model then focuses on the undulations of the thermocline in response to the winds. By changing the specified thermal structure of the ocean — which can be regarded as a background state — we learn how such changes affect El Niño. These simplified models are powerful tools with many uses, but they cannot tell us about the processes that maintain the background state itself. For example, the simple models are unable to tell us how global warming will affect the background state and hence El Niño. Scientists at a number of institutions worldwide are therefore developing models that simulate not only El Niño but the background state too. Several of these coupled General Circulation Models of the ocean and atmosphere, as they are known, reproduce southern oscillations that resemble the observed one in some respects, but at present they all have unrealistic features.[14] For example, the frequency with which El Niño appears varies considerably in these models, from every two years to every ten years, or never. The spatial structures of the simulated phenomena similarly cover a broad spectrum of possibilities. The models are improving rapidly and soon will be able to simulate the observed Southern Oscillation realistically. It will then be possible to use the models to predict El Niño and to explore how global warming will affect that phenomenon. To establish confidence in such predictions, the models will have to be tested. Reproducing the world of today is one test but is not sufficient to establish confidence in predictions concerning the future. It is of critical importance that the models succeed in simulating the different climates our planet has experienced in the distant past. To many people the study of past climates is a matter only of academic interest, but they are seriously mistaken.

SIXTEEN

Taking a Long-Term Geological View

This is the foundation of all: we are not to imagine or suppose, but to discover, what nature does or may be made to do.

—Francis Bacon (1561–1626)

Is the temperature rising yet? Is there evidence in the records for the past several decades that the current rapid increase in the atmospheric concentration of greenhouse gases is contributing to global warming? The debate about global warming focuses on these questions because of our penchant for doing nothing about a potential problem until it can no longer be denied. It is shortsighted, however, to assess the present only in the light of the recent past. Much can be learned about "what nature . . . may be made to do" by taking a longer-term view, one that only the geological record can provide. Such a perspective requires some familiarity with the remarkable past of the remarkable planet Earth.

Each of the planets in our solar system is a wondrous world — Saturn is adorned with innumerable rings, Jupiter with several moons — but only Earth is blessed with a miracle, a glorious diversity of flora and fauna. Only ours is a habitable planet. For several billion years temperatures on Earth have never approached the extremes of frigid Mars or hellish Venus. This planet has maintained conditions that favor life by engaging in a frenzy of activity. Earth spins about its tilted axis, orbits the Sun, moves its atmosphere and oceans in chaotic patterns, and punctuates the stately drift of its continents with sporadic quakes and eruptions.

Because of that drift the richly varied landscape we see today—frozen polar caps, hot, desolate deserts, lush jungles, spectacular mountains, and fertile prairies and pampas—is but a snapshot of a panorama that, over millions of years, has changed continually to the rhythms of cyclic phenomena such as the repeated buildup and erosion of mountains, the periodic opening and closing of ocean basins, and the disappearance and reappearance of huge glaciers. Let us confine our attention to the most recent era, the Cenozoic, during which mammals, birds, and flowers evolved. (It covers the past 60 million years, about 15 percent of the age of Earth.)

At the start of the Cenozoic the Atlantic Ocean was far more modest in size than it is today, the Pacific was even bigger than at present, Australia and South America were attached to Antarctica, and India was an island south of the equator.[1] At that time Earth was far warmer than it is today. The era of the dinosaurs had just come to an end, and temperatures at the poles were still sufficiently high for palm trees and crocodiles to thrive there. Earth had been very warm for tens of millions of years, but then very gradual global cooling commenced. Figure 16.1A shows how, over the past 60 million years, temperatures in high latitudes decreased from approximately 12° C to their present, freezing values.[2] This cooling was related to the appearance of various mountain ranges whose steep slopes promoted a process known as weathering, which removes carbon dioxide from the atmosphere, thus reducing the atmospheric greenhouse effect. The Himalayas appeared after India collided with Asia; the Andes and the Rockies emerged as the eastern floor of the Pacific plunged beneath the Americas. Figure 16.1A shows that the gradual global cooling that started 60 million years ago has continued right up to the present. Today the poles are covered with ice, but fossils provide evidence of palm trees in the distant past.

A period of very special interest to us, because its climate favored the evolution of our species,[3] is the past 3 million years. Eastern equatorial Africa used to be a hot, humid, tropical jungle but started becoming cooler and drier around 3 million years ago as a savanna landscape emerged. This transition was anything

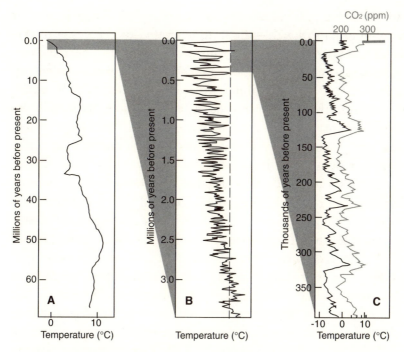

Figure 16.1. (A) The gradual decrease in surface temperatures in high latitudes over the past 60 million years, inferred from measurements in sediments deposited on the ocean floor. (B) A more detailed picture of the cooling over the past 3 million years. Fluctuations, mainly in global ice volume, are seen to grow in amplitude as we approach the present. (C) A detailed picture of the Ice Age cycle, over the past 400,000 years. The dark line shows variations in temperature and the light line shows variations in the atmospheric concentration of carbon dioxide, as determined from ice cores from Antarctica. Note the recent, very rapid rise in carbon dioxide levels since the start of the industrial revolution. For details see Zachos et al. and Petit et al. (chap. 16, notes 1 and 2).

but smooth and steady. It was marked by the appearance of continental glaciers in high northern latitudes, glaciers that started to wax and wane with an amplitude that increased steadily as we approach the present. This is clearly evident in the middle panel (B) of figure 16.1, which shows that we are living in a geological period of huge climate fluctuations. These oscillations, recurrent Ice Ages, have a very slow beat of many millennia. To find their origin we have to look beyond our planet, at our solar system.

Whereas the gradual cooling in figure 16.1A has a source of energy internal to Earth—the heat generated by the decay of certain radioactive elements deep inside our planet—the relatively rapid fluctuations in the middle panel (B) have a source of energy external to our planet. Milutin Milankovich identified the external source, in the nineteenth century, as periodic variations in the distribution of sunlight on Earth associated with the complex dance our planet performs as it orbits the Sun once a year while spinning daily about a tilted axis.[4] The daily and seasonal cycles are rhythms known to everyone, but astronomers are aware of additional, more subtle, and much slower beats, attributable to the gentle influence of the Moon and the other planets on Earth. For example, the tilt of Earth's axis, the reason for the seasons, rocks back and forth, between 22° and 24°, with a period of 41,000 years. For half that cycle the tilt increases, giving the poles more sunlight, the tropics less; then the tilt decreases. The axis also precesses, the way a spinning top does, causing a change in the distribution of sunlight with the seasons: warmer summers follow colder winters for about 11,000 years whereafter, for the same length of time, colder summers follow warmer winters. In addition, the orbit changes continuously from a nearly perfect circle to a slight ellipse and back to a circle every 100,000 years. This complex dance is choreographed very carefully, strictly in accord with Newton's laws. On the basis of those laws, and measurements of planetary orbits, astronomers can accurately reconstruct the dance over millions of years in the past. That is how they established that variations in sunlight have three slow beats at periods of 23,000 years, 41,000 years, and 100,000 years. It so happens that exactly the same periods characterize the fluctuations in the climate records shown in figure 16.1B. Such a remarkable coincidence cannot be accidental. We are obliged to conclude that the recurrent Ice Ages are in response to the periodic variations in sunlight. This does not amount to an explanation for the Ice Ages, however, because we still do not know why Earth's response to the same variations in sunlight is huge today but was very modest earlier than 3 million years ago. (As yet we have no explanation for the amplification of the cycle of Ice Ages shown in figure 16.1B.) What we can conclude from the geological record is that our species is thriving at a time in Earth's history when its climate is very sensitive to modest perturbations. By increasing atmospheric carbon dioxide levels rapidly, we are introducing an exponentially growing disturbance at a time when Earth's climate is far more sensitive to disturbances than at any other time during the Cenozoic.

From a geological perspective the present is unusual in another respect: figure 16.1C shows that the warm interglacial conditions we are currently enjoying occur infrequently. (Similar conditions last prevailed 130,000 years ago.) During such periods the atmospheric concentration of carbon dioxide, whose fluctuations are highly correlated with those in temperature, are at a maximum. This is part of the reason why, in a little more than a century, we have succeeded in raising carbon dioxide levels to values higher than they have been in more than 400,000 years. In figure 16.1B the amplitude of the cycle of Ice Ages is seen to increase as we approach the present. As a result interglacial conditions today are close to what they were some 3 million years ago. Is there a risk that the current rise in carbon dioxide levels could return us to the warm world of more than 3 million years ago? How did that world differ from the present one?

Today, the salient features of sea surface temperature patterns in low latitudes are the surprisingly cold surface waters in certain oceanic "upwelling" zones that include the eastern equatorial Pacific, the coastal zones of Northwest and Southwest Africa, and the coastal zones of California and Peru. In those regions, cold subsurface waters "well up" in response to the prevailing winds, except when those winds relax, in the eastern equatorial Pacific during El Niño for example. El Niño is at present an intermittent phenomenon, but in the geological records there is evidence that, earlier than 3 million years ago approximately, El Niño conditions were perennial.[5] Surface temperatures in the eastern equatorial Pacific have decreased over the past 3 million years while those in the west remained warm. The surface waters were persistently warm, not only in the upwelling zones of the equatorial Pacific, but in those off California and off Southwest Africa too.

In the light of this empirical information, the return of permanent El Niño conditions in response to the current rapid rise in the atmospheric concentration of carbon dioxide is within the realm of possibilities.[6]

Conditions today appear to be both close to and far from those of 3 million years ago. We are close to the earlier warm world because, today, the distribution of the continents and the composition of the atmosphere are only slightly different from what they were then. We are far from that world because El Niño is intermittent today but was perennial then. What are the factors that can favor an abrupt switch from the current conditions to those of the past? This important question is receiving considerable attention at present. Of special interest are the different mechanisms that can cause an increase in the sea surface temperatures of the eastern equatorial Pacific. Today, the warming of that region during El Niño involves an east-west redistribution of the warm surface waters without the net addition of heat to the ocean. Intense trade winds along the equator during La Niña drive the warm water westward and expose cold water to the surface in the east. When the winds relax during El Niño, the warm water flows back east, where the thermocline deepens. An alternative, less nimble way for warming the eastern Pacific is by adding heat to the ocean, thus increasing the volume of warm surface water and deepening the thermocline. Once the thermocline is so deep that the winds are unable to bring cold water to the surface, then El Niño conditions are permanent rather than intermittent. (From measurements in the western equatorial Pacific we learn that when the thermocline has a depth of approximately 150 meters, then its vertical excursions leave local surface temperatures unaffected.) Unlike familiar El Niño, which involves merely a horizontal redistribution of warm surface water in the Pacific, a permanent El Niño requires a net oceanic heat gain. This unfamiliar phenomenon, which scientists are only beginning to explore, clearly involves changes in the oceanic heat budget.

In a state of equilibrium, the ocean has a balanced heat budget so that the gain of heat equals the loss of heat. Today the gain is mainly in upwelling zones of low latitudes, such as the eastern equatorial Pacific, where cold water rises to the surface. The loss of heat occurs in higher latitudes, especially where cold air off the Asian and North American continents flows over the warm Kuroshio Current and Gulf Stream. Oceanic currents transport heat from the regions of gain to the regions of loss, thus ensuring a balanced heat budget for the ocean. Earlier than 3 million years ago, before the appearance of cold surface waters in the upwelling zones of low latitudes, the constraint of a balanced heat budget was probably satisfied in a different manner—locally everywhere. Hence the oceanic circulation must have been different, and the thermal structure of the ocean, which the circulation maintains, must also have been significantly different.

Up to now, oceanographers investigating abrupt changes in the oceanic heat transport to high latitudes have focused on the thermohaline circulation. Theoretical studies show that a freshening of the surface waters of the northern Atlantic could cause sudden changes in that component of the circulation, changes that could have a significant effect on the climate of northwestern Europe. How will changes in the salinity of the surface waters affect the other major component of the circulation, the wind-driven gyres? Attention is now turning to this question. It appears that, as in the case of the thermohaline circulation, so the wind-driven circulation can be subject to abrupt changes, except that the changes manifest themselves differently. They become apparent, not in the northern Atlantic, but in the tropics, in the form of perennial El Niño conditions. These tentative results suggest that global warming, should it cause the melting of the polar ice caps and hence a freshening of the surface water in high latitudes, could result in a permanent El Niño. The events of 3 million years are valuable tests for these theories.

Some 3 million years ago La Niña made an entrance, bringing to an end an era of permanent El Niño conditions. The associated climate changes—our planet became more temperate—favored the rapid rise of our species. Within a remarkably short time we acquired the ability to change the atmospheric composition. How ironic it would be if our activities result in the banishment of La Niña, and thus the termination of the very conditions that are allowing us to flourish.

Part 5 | Coping with Hazards

SEVENTEEN

Famines
in India

Science has the first word on everything, the last word on nothing.
—Victor Hugo (1802–1885)

During the latter half of the nineteenth century horrendous famines in India resulted in the death of millions of people.[1] Some believed that nothing could be done, because the root of the problem was overpopulation: "every benevolent attempt made to mitigate the effects of famine and defective sanitation serves but to enhance the evils of overpopulation."[2] Others believed that nothing should be done, because "famine has never arisen from any other cause but the violence of government attempting, by improper means, to remedy the inconvenience of dearth."[3] These followers of Adam Smith argued that the market would solve the problem, that high food prices would promote imports and limit consumption. Almost everyone assumed that famine is the result of a lack of food, which in turn is a consequence of poor rainfall. "Indra and Vayu, the watery atmosphere and the Wind, are still the prime dispensers of weal or woe to the Indian races."[4] This led to the assumption that the prediction of droughts is the key to mitigating the horrors of famine. To that end, the 1877 Famine Commission asked Henry Blandford, first director of the India Meteorological Department, to develop methods for forecasting drought. His methods had some initial success but failed in the late 1890s, when India again suffered a ghastly famine. Early in the twentieth century Gilbert Walker arrived in India to continue the development of prediction methods. He found that poor monsoon rains tend to occur during one phase of the Southern Oscillation, the phase that corresponds to El Niño, but he was unable to convert that result into a predictor of droughts.

Over the past century food production in India has increased substantially, but so has the population; there has been relatively little change in per capita consumption of food. India still experiences rainfall fluctuations; occasionally droughts still cause a significant reduction in the food supply. From 1970 to 1973 the state of Maharashtra suffered a prolonged drought that reduced food production by a factor of two; in 1987 and 1988 monsoons were again poor. However, for several decades now, there have been no famines in India. Are the successors of Blandford and Walker producing accurate scientific forecasts of rainfall fluctuations to facilitate planning and policy implementation? Although we have learned much about climate fluctuations in general, El Niño in particular, the accurate prediction of year-to-year variations in Indian rainfall remains an elusive goal. India copes with droughts without relying on accurate scientific forecasts.

A natural hazard—drought, flood, etcetera—can affect a society in two ways, by destroying crops and by affecting adversely the economy associated with agriculture.[5] For example, those who provide goods and services can lose their employment and hence income. If large numbers of people are unable to purchase food, a famine is possible even when the supply of food is adequate. A transportation system for distributing food, and markets to make the food available to people, do not necessarily solve the problem. (Transportation can even exacerbate the problem; during the Irish famines of the 1840s food was shipped from Ireland to England, where consumers could pay high prices.)

The successful strategy adopted in Maharashtra in 1970–1973, when food production was nearly halved, had several aspects. The local government started a massive public works program that provided large numbers of people with cash to purchase food. The demand enabled private traders to import and distribute food. To inhibit a rise in prices, the government sold reserve stocks that had been accumulated during years of plenty. A potential catastrophe was avoided, not only because a transportation system and markets existed, but because people had the means to buy food. Perhaps the most important factor was a government willing, and able, to act. The Famine Commission, already in the 1880s, had identified measures to mitigate the impact of droughts, but for a long time politicians largely ignored those. In a democracy, elected officials pay a price for failing to solve recurrent social problems. Once India became an independent country with elected officials, progress in coping with droughts was rapid. Famine is a scourge that continues to plague some countries where the "absence of economic power combined with a lack of political leverage [has] condemned millions of people to unrelieved destitution and untimely death."[6] Democracy can contribute to the solution of social problems such as famines but has thus far failed to cope with several other serious problems. Starvation, for example, persists, even in some democracies.[7]

Reliable forecasts of El Niño, and more generally of climate fluctuations and climate changes, can be of enormous value in our attempts to cope with the associated droughts, floods, and other disruptions. However, the prevention of disasters depends critically on political factors. In some poor countries, those uninterested in political change may find it convenient to identify El Niño as a culprit responsible for droughts and social problems. Scientific results can be used for unintended purposes.

EIGHTEEN

Fisheries of Peru

To everything there is a season,
  and a time to every purpose under the heaven;
A time to be born, and a time to die;
  a time to plant and a time to pluck up that which is
  planted;
—Ecclesiastes 3:1

Peru is blessed with the richest fishing grounds on Earth because it has a coast that stretches across a band of low latitudes and has strong winds parallel to that coast. The winds drive coastal currents that veer offshore, and because the thermocline happens to be very shallow, cold water, rich in nutrients, wells up from below. The aquatic environment off Peru — surface waters with plentiful sunlight and nutrients — is ideal for phytoplankton, the marine plants at the base of the food chain.

The coast of Peru has another unusual feature, large seasonal variations in temperature despite minimal seasonal variations in sunlight. The waters are coldest in September and October, when the winds are strongest; the opposite phase of the seasonal cycle is in March and April. These annual fluctuations, and the superimposed cycle between La Niña and El Niño, result in temperature variations that have a significant effect on the various life-forms. For example, the appearance of warm water in the eastern tropical Pacific during El Niño is beneficial to tropical species that prefer high temperatures, but the Peruvian anchovy, which thrives when temperatures are in the range from 14.5° C to 21° C, has difficulties reproducing when temperatures exceed 21° C.[1] An intense El Niño, when exceptionally warm waters persist off

Peru and Ecuador for an unusually long time, can be devastating to anchovies, but it usually takes those fish only a few years to recover. Because their life span is four to six years, failure to reproduce for a year or two is a setback they readily overcome. Their size, 17 centimeters approximately, also permits the fish to swim to more favorable, colder conditions when warm waters appear. (It is thought that, during El Niño, some of the fish off Peru migrate southward, to the colder waters off Chile.)

Earth's fauna and flora have come to terms with the seasonal cycle, and with El Niño and La Niña. The fish in the waters off Peru, and the birds that live off the fish, welcome La Niña as a time to be born, recognize El Niño as a time to die. Even the plants in the adjacent desert flourish during El Niño, wither during La Niña. To them the irregular, whimsical Southern Oscillation is as much part of the rhythm of life as is the more predictable cycle of summer and winter. Fish and fowl have learned to cope, but we humans are having trouble accepting that "to everything there is a season." (Fisheries are in trouble, not only off Peru, but around the globe.)[2]

The growth of highly organized fisheries in Peru started after World War II, after the collapse of the California sardine-fishing industry. Peruvians catch mostly small pelagics — primarily anchovies — to produce fish meal for animal feed in the United States and elsewhere. In 1954 Peru had some 16 plants to process the fish caught from slightly more than a hundred small boats of about 60 tons. Ten years later there were 150 plants to process the fish caught from close to two thousand boats with an average weight of about 350 tons, many equipped with sophisticated equipment (echo sounders, radar, automatic pulleys). This growth was so rapid that the fish population started showing signs of stress. The government of Peru therefore imposed regulations: in 1965 fishing was restricted to a five-day (Monday to Friday) week, and at least 50 percent of the catch had to consist of fish more than 12 centimeters in length. The 297 fishing days allowed in 1966 were reduced to 180 in 1970. These regulations did not limit the amount of fish being caught. Larger and faster boats permitted the catch to continue increasing sharply. Then in 1972, the year of an intense El Niño, there was a crash.[3] Almost no fish were caught. Was El Niño the culprit?

For fisheries to be sustainable, only a fraction of the fish population should be caught each year. Determining what that fraction should be is no trivial matter. Prudent management of fisheries is difficult anywhere; at present fishermen in eastern Canada and the northeastern United States are coping very poorly. Peruvians face an even bigger challenge because there the fish stocks are subject to large seasonal and interannual variations. Up to 1970 the catch increased rapidly, but then came the crash (see figure 18.1). After the crash in 1972 the fish stock remained depleted for more than a decade. The economic consequences were severe. At the end of 1975 there were only 530 boats and 51 factories in operation. Several factors contributed to this calamity. An obvious one is overfishing, the exploitation of a resource beyond the point where it can renew itself. However, it is also possible for populations to fluctuate even in the absence of human intervention. No species lives in isolation. Each has predators and prey, and competes with other species for food. As a consequence, the fortunes of a population can vary considerably as it gains and loses supremacy over other species. A modest climatic change that the eastern equatorial Pacific experienced during the second half of the 1970s—the trade winds weakened slightly, and the thermocline deepened—could have favored some species, disadvantaged others.

Several factors, some to a greater degree than others, caused the fish stock off Peru to remain low for more than a decade after 1972, but the stock started to rise again in the early 1990s. Measures introduced by the government of Peru to facilitate the recovery include regulations that reduce fishing during the spawning period of the anchoveta and protection of the juveniles until they reach maturity. A decision to put additional regulations into effect is usually based on estimates of the fish population. The size of the catch is one indicator. The fish catch remained large in April and May 1997, even though El Niño was known to be developing. Did that imply a large fish population and hence no need for special regulations?

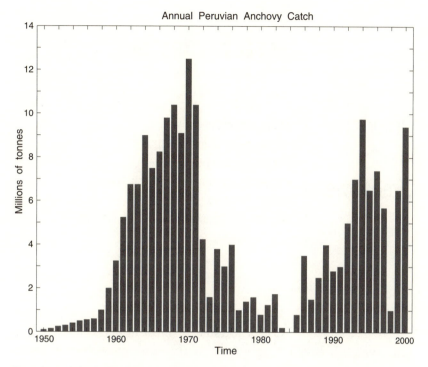

Figure 18.1. The annual Peruvian anchovy catch in millions of metric tons.

In 1997, information about the developing El Niño came mainly from satellite measurements of sea surface temperatures. That same information facilitated the catching of fish. As the warm water moved eastward across the Pacific and approached the coast of South America, cold surface water was increasingly confined to small pockets close to the shore. Schools of anchovies accumulated there and could easily be caught, provided that the pockets could be located. The satellite photographs provided that information. Hence it was incorrect to regard the size of the catch as a measure of the total population. Once this reason for the higher catch was recognized, the government banned all fishing until El Niño conditions retreated. After El Niño came to an end, the catch increased, and by the year 2000 levels were comparable to those of 1970. Is this a sustainable level? Time will tell.

NINETEEN

Droughts in Zimbabwe

the race is not to the swift,
nor the battle to the strong,
neither yet bread to the wise,
nor yet riches to men of understanding,
nor yet favour to men of skill;
but time and chance happeneth to them all.
　　　—Ecclesiastes 9:11

Chris Karsten, Princeton class of '97, carefully watches Bob Olango demonstrate the mysterious Coriolis force to a group of tourists on an excursion to the equator, just north of Nairobi, the capital of Kenya.[1] The well-educated tourists know all about the Coriolis force, how it causes the water in a draining toilet to spiral anticlockwise in the Northern Hemisphere, clockwise in the Southern Hemisphere. To demonstrate this effect, Bob Olango uses a conical bucket. Precisely 20 yards north of the equator he fills the bucket with water, floats three matchsticks on the surface to visualize the motion of the water, and pulls the plug. The tourists lean forward and observe that the motion is indeed anticlockwise. Everybody then follows Bob as he carefully paces off 20 yards to the equator, plus another 20 into the Southern Hemisphere. There Bob repeats the demonstration. This time the motion is clockwise, as expected. Everybody nods approvingly. To the educated, it is reassuring that they "know" the reason for this esoteric natural phenomenon. Confirmation that they benefited from the "physics for poets" course in college many years ago is most welcome. They ask Bob to pose for a photograph with them and eagerly pay him $10 for a formal document, which states that, on such and such a day and year, the bearer witnessed a demonstration of the Coriolis force in Nairobi, Kenya. In London and Chicago, friends and acquaintances will proudly be shown this document, and everyone will nod approvingly. Very few will realize that Bob's demonstration is a hoax, that Bob cleverly exploits the tourists' tenuous grasp of science.

The "demonstration" is so convincing that Chris Karsten meets with skepticism when he tries to explain to the tourists that, yes, there is a Coriolis force in reality but, no, it has nothing to do with what happens in Bob's bucket. (Chris has benefited from his geosciences courses at Princeton.) The tourists believe what they have seen with their own eyes and dismiss Chris's arguments. He gradually comes to the conclusion that Bob Olango is a most impressive and ingenious entrepreneur.

Bob grew up near the equator and for many years never realized that there is something special about that line. Then he gradually became aware that, to the steady stream of tourists, the line is special. After learning about the Coriolis force from a chatty visitor, Bob set himself the challenge of capitalizing on the foreigners' belief in this mysterious force. He saw a business opportunity. He experimented with various buckets and at first found that, at any location, the water sometimes swirls out one way, sometimes the opposite way. Only after some practice — an imperceptible flick of the wrist — was he able to make the water swirl consistently clockwise or, if he so chose, anticlockwise. The location of the equator does not matter in the least — Bob can perform his demonstration equally well on the opposite sides of an arbitrary line in London or Chicago — but, for obvious reasons, Bob chooses to perform across the special line just north of Nairobi. That is where his customers wish to see the demonstration. By observing his spectators closely, Bob learned how to polish his performance. After identifying the symbols of authority that the foreigners respect, he bolstered their confidence in his demonstration of the mysterious Coriolis force by enhancing his air of authority. He donned an official-looking blazer (discarded by a member of the local cricket club) and persuaded an acquaintance at a printing shop to produce the "official" certificates that he sells. These innovations led to a significant growth in Bob's business.

A fascination with the equator is common among the educated classes. On cruises that cross that line, the occasion calls for a festive, amusing ceremony to initiate those who are visiting the equator for the first time. The humor stems from the pretense that the line being crossed is special, when everybody knows that the line is of no great import (except for the effect on the water swirling in a toilet). But in reality the line is very special. It is associated with many curious oceanic phenomena that arise because, on a rotating planet, the Coriolis force deflects moving particles toward their right in the Northern Hemisphere, their left in the Southern Hemisphere. (The Coriolis force vanishes along the equator.) Artillery officers are aware of this and, in the Northern Hemisphere, fire shells slightly to the left of the target, counting on the Coriolis force to deflect the shell to the right. The same force causes a parting of the waters at the equator when the winds are driving those waters westward. The Coriolis force pushes water parcels northward if they happen to be north of the equator, southward if they are south of the equator. As a result, the equator is a line from which the surface waters diverge so that cold water, rich in nutrients, rises to the surface from below. That is why the equator can easily be identified in satellite photographs of sea surface temperatures; it is at the center of a tongue of very cold surface waters in the Pacific. In photographs of chlorophyll concentration in the ocean, very high values mark the location of the equator because it has an abundance of sea plants (phytoplankton).

The parting of the waters along the equator depends mainly on two forces; the westward winds provide one, the Coriolis force the other. In general, a parcel in motion on the surface of the earth is subject to several forces. When water spirals out of a toilet, gravity accelerates it downward while frictional forces retard it. The Coriolis force too has an influence, but given the small size of the toilet, that force is minuscule in comparison with the others. To appreciate why size matters, keep in mind that the

Coriolis force depends on the rotation of the earth, once a day about its axis. A parcel of air spiraling in a hurricane is strongly influenced by the Coriolis force because that parcel travels huge distances over prolonged periods of many hours, during which the rotation of the earth is significant. Because of the Coriolis force, motion in a hurricane is anticlockwise in the Northern Hemisphere, clockwise in the Southern Hemisphere. A tiny dust devil on a mesa in Arizona lasts for only a few seconds and is essentially unaffected by the rotation of the earth. It can spiral either clockwise or anticlockwise. A shell fired from a cannon travels for a brief period during which the earth hardly rotates, so the Coriolis force has a small effect. (That effect is nonetheless important to an artillery officer trying to hit a specific building miles away, not the hospital next to it.) In a toilet bowl with dimensions that can be measured in inches, the Coriolis force has an even smaller effect and is overwhelmed by the other forces that Bob Olango cleverly brings into play when he pulls the plug from his bucket.

After watching the demonstration at the equator, Chris concluded, not for the first time on his travels around Africa, that the continent has no lack of smart, talented people. What Africans need are opportunities, but what they are offered is advice. Chris too was in Africa to give advice, of a scientific nature. He would soon have reason to wonder whether there are parallels between his and Bob Olango's use of science.

When he graduated from Princeton, Chris Karsten declined several enticing jobs, mostly on Wall Street, and accepted a position with NOAA, a government agency in Washington, D.C., responsible for weather and climate forecasts. Chris found himself in the right place at the right time. El Niño of 1997 was in full swing when he arrived at NOAA, and the demand for detailed information exploded. The motives of the callers varied. Relief agencies wanted information for obvious reasons. Officials from the State Department were concerned about the stability of some of the countries in southeastern Asia and South America. The droughts and floods that El Niño induces in different regions contribute to forest fires and other disasters that can topple governments. When would the drought in southeastern Asia be over? When does the rainy season in that part of the world usually start? Chris's advice was, and continues to be, much in demand. His job is to convey scientific information from the producers to the users of that information. Often the youngster finds himself teaching elementary geography—the seasonal cycles in different parts of the world—to experts on the politics of those different parts of the world.

In late 1997 El Niño was unleashing torrential rains on Kenya and was threatening to inflict droughts on Zimbabwe to the south. Chris started traveling to that part of the world to assist with relief efforts. On one of those trips he saw Bob Olango's demonstration at the equator. On the same trip he then continued to Zimbabwe, where he met Nelson Mgabe, who in due course impressed Chris even more than Bob did.

Nelson Mgabe is a farmer who has endured and survived numerous floods and droughts. He first heard of El Niño at a gathering in Harari, where his government was alerting farmers of an imminent drought. Chris Karsten was one of the foreign scientists and advisers who explained how oceanic conditions in the far eastern Pacific, on the other side of the globe, were about to interfere with the rains over Zimbabwe. The farmers were advised to prepare by purchasing drought-resistant seeds. Several asked questions about this El Niño and his astonishing powers. Nelson wanted to know whether other farmers in Zimbabwe were also being told of the imminent drought. He was assured that everybody was being told.

More than a year later, after La Niña has replaced El Niño, Chris happens to be back in Harari. He is aware that the scientific forecasts had been inaccurate, that Zimbabwe enjoyed normal rainfall despite the presence of an intense El Niño in the Pacific. Normal rains are surely better than poor rains, so Chris assumes that the inaccurate forecast was a matter of no consequence. He is dismayed to learn that, because of the forecast, crop production in Zimbabwe was 20 percent lower than normal. Apparently the banks paid close attention to the weather prophets from overseas and declined loans to farmers on the grounds that the farmers would be unable to pay back the loans. One person, at least, did emerge unscathed. That was Nelson.

Over the years Nelson has been the recipient of advice from a succession of foreigners — representatives of the Church, of the Crown, of the Bank, and most recently of Science. He nonetheless learned how to cope with droughts and other hardships. When he was told of distant El Niño's purported ability to influence rains over Zimbabwe, he was skeptical and decided that, as in the past, he would be ready for whatever happened. He anticipated that a demand for drought-resistant seeds would cause prices to rise steeply. He refrained from getting any but did save part of the crop of the previous year, as usual. He planted as usual.

"Time and chance happeneth to them all."

EPILOGUE

Becoming Custodians of Planet Earth

Constant El Niño could soon become fickle. This is disturbing news because we are having more and more trouble coping with familiar El Niño. Will he grow more intense? Will his brief visits become prolonged? Could the sporadic visitor turn into a permanent resident? As yet we have no definite answers, but we do know that a change in the properties of El Niño is inevitable should the atmospheric concentration of greenhouse gases continue to rise. We are rapidly changing the composition of the atmosphere, not by design, but as an unfortunate by-product of industrial and agricultural activities that bring us enormous benefits—increasing standards of living for the rich and poor alike. For how long will these considerable benefits outweigh the possible adverse consequences of global climate changes that include an altered El Niño? We are gambling that, for the time being, the odds are in our favor. At what stage will the risks become unacceptably high? In a well-known song Kenny Rogers reminds us that gamblers "got to know when to hold 'em, know when to fold 'em."[1] Do we?

Some people are convinced that it is far too early to "fold 'em." They note that our winnings, thus far, have been handsome and that the beneficiaries include many poor nations. Given that policies to mitigate the potential impact of global warming could harm the economies of many countries, and given the considerable uncertainties in the scientific results concerning global warming, is it not prudent to defer action until we know exactly how the rise in the atmospheric concentration of greenhouse gases will affect our climate? Is global warming such a serious threat that it merits immediate action?

This line of reasoning is persuasive to people who find the term global warming vague and not particularly threatening. (A few years ago, after an exceptionally hot summer in Europe, a number of Germans were asked what action they would favor should global warming cause such hot summers to become more frequent. Some proposed longer summer vacations. Other wished for more air conditioners.) The impression that global warming is not an urgent problem is at first reinforced when people learn that the problem will be serious, not for our children, but for their great-grandchildren and subsequent generations. Such a nonchalant attitude toward the calamities that will be visited upon future generations changes once it is recognized that those calamities will be attributable to our activities today. The decisions we make today will affect, for many generations to come, not only humans, but all the inhabitants of this planet.

Over the past century we have made astounding technological advances and have become so powerful that we now are geologic agents, capable of interfering with the processes that make Earth a habitable planet. By the end of the nineteenth century scientists had already hypothesized that our industrial activities would increase the atmospheric concentration of greenhouse gases and thus cause global climate changes. It took several more decades to establish that we humans are indeed in the process of significantly altering the composition of the atmosphere. That important scientific result implies that the power we now wield is so enormous that we have become custodians of planet Earth. Some extreme environmentalists are convinced that the best way to proceed is to trust nature because it knows best. They insist that we should desist from any attempts to reorganize nature. For example, we should refrain from burning fossil fuels and increasing the atmospheric concentration of carbon dioxide. Theirs is an untenable position because we too are part of nature and hence have the right to take actions that ensure our survival and prosperity. However, in doing so, we must keep in mind that, as custodians of this planet, we have responsibilities toward all its inhabitants and have to act on their behalf.

In making decisions about how to conduct our affairs as wise and responsible stewards of this planet, we should seek guidance from scientists, engineers, and economists concerning matters such as global warming—how and when it will affect climate in different parts of the globe, the likely cost of its impact, and possible methods to mitigate that impact. However, the final responsibility for the welfare of the planet rests with all of us. The matter is too important to be left to technical experts. Everyone has to participate in determining our responsibilities toward future generations and toward those who cannot participate in our debates.

To be responsible custodians of planet Earth, all of us should have some familiarity with its history and should have a rudimentary understanding of the processes that permit a great diversity of species to flourish. Earth is so complex that this is a serious challenge. It is therefore heartening that, over the past century, we have made enormous progress in coping with a system that is even more complex, the human body. We have increased life expectancy by several decades, mainly because we demand that every educated person should be familiar with the workings of the human body—the functions of the heart, liver, lungs, etcetera—and should know about the importance of a healthy diet, exercise, and hygiene. We firmly believe that prevention is better than cure, and we know which activities contribute to the prevention of diseases. Everyone accepts that the health of a population depends on far more than excellent medical facilities. Each of us needs to become as informed about planet Earth as we are about our bodies. Particularly important is awareness of the effect of our daily activities on the environment. To live in harmony with nature, a passionate expression of concern about the environment is no substitute for some familiarity with the processes that make this a habitable planet.

Ours is a very special planet, the only one known to be habitable.[2] A surprisingly large number of fortuitous factors contribute to this state of affairs. Earth is at just the right distance from the center of our galaxy, is at the right distance from our sun, has the right size, has the right atmospheric composition, has the right surface area covered with water. . . . So many factors have to be just right for life to be possible that preciously few planets in the universe are likely to be habitable. Not only do we live on a special planet, but we are doing so at a special time in its long history (a matter discussed in chapter 16). For the past million years the salient features of Earth's climate have been the recurrent Ice Ages. Relatively warm periods such as the one we are experiencing at present occur infrequently and are separated by prolonged periods (close to 100,000 years) during which glaciers cover large parts of the continents. The current warm, interglacial epoch, which favored the development of agriculture and the growth of civilizations, started some 10,000 years ago. The onset of these warm conditions was accompanied by a rise in the atmospheric concentration of carbon dioxide, a rise that contributed to the increase in temperatures (see figure 16.1). When our industrial activities started 150 years ago, carbon dioxide levels were already at a natural maximum, the reason why the atmospheric concentration of that gas today is the highest it has been in more than 400,000 years. Given that another Ice Age is imminent, the current increase in carbon dioxide levels may seem desirable. The problem, however, is that we are increasing those levels at much too rapid a rate. (Before the end of this century the atmospheric concentration of carbon dioxide is likely to be twice what it was when the industrial revolution started.) The consequences of our activities will be evident long before the onset of the next Ice Age becomes apparent over the course of a few millennia. The following riddle helps explain why the current rapid increase in carbon dioxide levels is cause for serious concern.

A gardener one day finds a beautiful lily in his pond. The next day there are two, the day thereafter four, and so on. The number of lilies in the pond doubles every day. The gardener soon realizes that, if left unchecked, the lilies in due course will choke everything else in the pond. If the lilies would cover the pond completely 100 days after their first appearance, on what day should the gardener take action? It may seem reasonable to wait until the pond is half full of lilies. However, to do so is to invite disaster because that happens on day 99. To wait until a quarter of the pond is covered with lilies buys the gardener almost no additional time because that happens on day 98! Suppose that the gardener quickly digs another pond, comparable in size to the original one, and transfers half the lilies to the new pond. How much time does he buy? On what day will both ponds be covered with lilies? The answer is day 101! The lesson this riddle teaches is that matters can get out of hand very suddenly when dealing with exponential growth.

Although Albert Einstein once described compound interest as "the greatest mathematical discovery of all time," few people appreciate the implications of compound (or exponential) growth. Consider the current rapid rise in the atmospheric concentration of greenhouse gases. To debate whether or not the consequences of that exponential growth are already discernible is equivalent to debating whether, in the case of the lilies in the pond, the gardener is at day 80 or day 90. Far more important than a precise answer that settles the issue is recognition that, when dealing with exponential growth, it is far better to act sooner than later.

We face a potentially serious environmental problem that requires difficult policy decisions. Uncertainties in the scientific results complicate matters enormously. What guidance is available from our long history of dealing with environmental problems on the basis of uncertain scientific information? Past experience with severe storms, hurricanes, prolonged droughts, and so forth tells us that uncertain information can be of considerable value. Weather forecasts, long before they were as accurate as they are today, saved many lives and could have saved even more if the captains of boats off New England, and admirals of fleets in the Pacific, had paid closer attention to those forecasts. Uncertain information does of course have limitations. The available scientific results concerning global warming leave no doubt that globally averaged temperatures will rise should the atmospheric concentration of greenhouse gases continue to increase, but the results are unreliable as regards the exact nature of the hazards — droughts, floods, rising sea level, etcetera — that a particular region will have to endure. Apparently some people are unable to respond to the seriousness of our potential problems until they know exactly how their specific district will be affected. Such seemingly detailed information can be obtained by extrapolating from the available climate forecasts. However, the uncertainties in such extrapolations are so large that, at present, the results are of little value. In due course scientific advances will improve this unsatisfactory state of affairs. In the meanwhile we must keep in mind Maynard Keynes's remark that it is better to be vaguely right than to be precisely wrong. The available scientific information has uncertainties, but is sufficient to justify action.

The cause of our potential problems is the rapid increase in the atmospheric concentration of greenhouse gases. Our goal should therefore be a reduction in the rate of emission of greenhouse gases into the atmosphere. Because the growth is exponential at present, merely reducing the rate at which we burn fossil fuels can amount to a significant contribution. In the earlier example of exponential growth, in which the lilies in a pond double every day, reducing the rate of growth below 100 percent per day would benefit the gardener enormously, but that option is not readily available to him. We, however, in our consumption of fossil fuels, can easily reap large benefits by merely reducing the rate at which our use of oil, coal, and other fossil fuels is increasing. Not only can we delay the onset of global warming, but we can also temporarily allay fears that our limited supply of oil will be exhausted in the near future. By merely reducing the rate at which consumption grows, from 5 to 2.5 percent annually, say, we can almost double the life expectancy of the current supply.[3] Much more can be achieved by being more ambitious. Supplies that are depleted after merely 36 years when the consumption grows at 5 percent per year last for 100 years if the rate of consumption remains constant (so that the rate of growth is zero).

Supplies that are depleted after 79 years if the demand grows at 5 percent per year last for 1,000 years with steady consumption. Supplies that are available for merely 125 years if the usage grows at 5 percent per year last 10,000 years with steady consumption. A far better strategy than searching for more reserves is to reduce the rate at which our consumption grows. Those who insist that conservation is an ineffective way to deal with a potential energy crisis, who insist that the only solution is to find more reserves, are woefully ignorant of the implications of compound interest. Greater efficiency will make the limited supply of fuels last longer, will make us less dependent on imports from other countries, will reduce the rate at which we emit greenhouse gases into the atmosphere, and hence will delay the onset of global climate changes. These simple arguments imply that there are indeed very effective ways in which everyone can contribute to mitigating the potential impact of global warming: everyone has to become more efficient in the use of energy derived from the burning of fossil fuels. In the United States, several large corporations and companies are already taking unilateral steps to be more efficient in their use of energy.[4] Their example needs to be followed widely.

Individuals can make important contributions, but ultimately stabilization of the composition of the atmosphere has to be an international effort because we all share the atmosphere. Our engineers and scientists are rapidly developing the means to reduce the emission of greenhouse gases into the atmosphere. (When encouraged to do so, our engineers are remarkably clever at solving certain problems by inventing new technologies.) Economists are proposing a number of strategies for a transition to the technologies being invented without causing significant harm to the economy. Presumably some of the proposals are efficient and effective. How do we identify them? By a method of trial and error, by putting some of the proposed solutions into practice. This has to be done in a flexible manner so that we can modify or abandon an approach that proves ineffective. We must steer clear of grand and ambitious proposals that purport to solve this problem once and for all by following a rigid course of action. We must also guard against the temptation to defer difficult political decisions because of a perceived need for more accurate scientific information. Such information will be of enormous value, is a high priority for scientists, and will be available in due course. In the meanwhile we can act on the basis of what we have already learned from our exhilarating and rewarding affair with El Niño. Presumably we all want him to remain steadfast. In that case we have to accept that the time has come "to hold 'em," and "to fold 'em."

NOTES AND REFERENCES

Prologue: Assessing Our Affair as It Approaches a Critical Juncture

1. R. A. Pielke, Jr., and R. A. Pielke, Sr., in *Hurricanes, Climate, and Socioeconomic Impacts*, H. F. Diaz and R. S. Pulwarty, eds. (Berlin: Springer-Verlag, 1997), 147–84. See also van der Vink et al., "Why the United States Is Becoming More Vulnerable to Natural Disasters," *Eos* (Trans. Am. Geophys. Union) 79, no. 44 (November 3, 1988), 533–37.

2. I learned about this from Marshall Mdoka at a meeting in Trieste, Italy, in June 2002. In 1997 scientists alerted not only California and Zimbabwe but several other countries of the likely impact of El Niño on their climate. Australians and Indians were told to expect droughts, but rainfall was normal in both cases. Indians paid attention to their own scientists, who predicted normal monsoons and proved correct. See page 108 of *Nature* 388 July 10, 1997.

3. James Trefil, *1001 Things Everybody Should Know about Science* (London: Cassell, 1993).

4. My notes for this course — attempts to explain the science of weather and climate to laymen — are summarized in *Is the Temperature Rising?* (Princeton, N.J.: Princeton Univ. Press, 1998). Some of the topics discussed in that book are revisited in this book.

Chapter 1. A Mercurial Character

1. Señor Federico Alfonso Perez made the following remarks in his address to the Sixth International Geographical Congress in Lima, Peru, in 1895:

In the year 1891, Señor Dr. Luis Carranza, President of the Lima Geographical Society, contributed a small article to the Bulletin of that Society, calling attention to the fact that a counter-current flowing from north to south had been observed between the ports of Paita and Pacasmayo.

The Paita sailors, who frequently navigate along the coast in small craft, name this counter-current the current of "El Niño" (the Child

Jesus) because it has been observed to appear immediately after Christmas.

As this counter-current has been noticed on different occasions, and its appearance along the Peruvian coast has been concurrent with rains in latitudes where it seldom if ever rains to any great extent, I wish, on the present occasion, to call the attention of the distinguished geographers here assembled to this phenomenon, which exercises, undoubtedly, a very great influence over the climatic conditions of that part of the world.

2. K. E. Trenberth, "The Definition of El Niño," *Bull. Am. Met. Soc.* 78 (1997): 2771–77.

3. SCOR (Scientific Committee for Ocean Research of IOC, UNESCO) Working Group 55, *Prediction of El Niño*, proc. No. 19 (Paris, 1983).

4. J. Tuzo Wilson, *IGY: The Year of the New Moons* (New York: Alfred A. Knopf, 1961).

5. For references and a more detailed summary of the scientific results, see part 4 of this book.

6. The term La Niña, to denote the complement of El Niño, first appeared in "El Niño and La Niña," an article by S.G. Philander in *J. Atm. Sci.* 42 (1985): 2652–62.

7. Stephen Jay Gould, *Time's Arrow, Time's Cycle: Myth and Metaphor in the Discovery of Geological Time.* (Cambridge, Mass.: Harvard Univ. Press, 1987).

8. J. Hutton, "Theory of the earth," *Transactions of the Royal Society of Edinburgh* 1 (1788): 209 305.

9. Difficulties with the term ENSO become apparent when authors, sometimes in the same paragraph, write of (1) ENSO events, (2) the ENSO cycle, (3) ENSO predictability, (4) the ENSO phenomenon, and (5) ENSO theory. From a close reading of the texts it is sometimes possible to divine that in (1) ENSO is none other than El Niño but that, in (2), the same term refers to the Southern Oscillation. In (3) ENSO probably refers to El Niño because, thus far, there is little interest in predicting La Niña. The "phenomenon" in (4) is at first a mystery but usually turns out to be either El Niño, La Niña, or the Southern Oscillation. In (5) ENSO refers to none of the possibilities mentioned thus far but to interactions between the ocean and atmosphere.

10. The decadal modulation of the Southern Oscillation, which as yet is unexplained and is a topic of considerable debate today, is part of the reason why the pioneering studies of G. T. Walker and E. W. Bliss ("World Weather," V, *Mem. Roy. Meteorol. Soc.* 4 [1932]: 53–84) concerning the Southern Oscillation fell into oblivion shortly after the results were published: the Southern Oscillation was energetic during the early decades of

the twentieth century but then faded away! An index for the oscillation, the correlation between surface pressure fluctuations at Darwin and Honolulu, had a value of −.66 before the 1920s but then fell to −.12 during the subsequent decades (Trenberth and Shea, "On the Evolution of the Southern Oscillation," *Mon. Wea. Rev.* 115 [1987]: 3078–96). By the 1960s some people started wondering whether the Southern Oscillation was an artifact of Walker's analyses of relatively short records. From the 1960s onward, the value of the Southern Oscillation became energetic again, and the value of its index increased again. The apparent change in the properties of El Niño in the late 1970s, evident in figure 1.1, has become a matter of much debate. Some investigators believe that the record is a stationary time series characterized by a distinctive timescale, plus noise that extends to periods of decades and more. Others propose that the decadal and longer-term fluctuations involve physical processes distinct from those that characterize the Southern Oscillation, and which modulate the properties of the latter oscillation. For a discussion, see A. Fedorov and S. G. Philander, "Is El Niño Changing?" *Science* 288 (2001): 1997–2002.

11. Figures that indicate how El Niño conditions in the tropical Pacific tend to affect different parts of the globe during the winter of the northern hemisphere can be found on websites such as *www.pmel.noaa.gov/toga-tao/*. Such figures can be misinterpreted unless the viewer takes into account that the probability that a certain region will be affected in the indicated manner varies considerably. The probability that El Niño will bring heavy rains to Peru is very high, but the probability for dry conditions in Zimbabwe, though statistically significant, is low.

12. La Niña conditions can appear in the Indian Ocean even in the absence of El Niño from the Pacific. See N. H. Saji, B. N. Goswami, P. N. Vinayachandran, and T. Yamagata, "A Dipole Mode in the Tropical Indian Ocean," *Nature* 401, no. 6751: 360–63.

Chapter 2. A Fallen Angel?

1. These quotations are from letters written by Mr. S. M. Scott of Florence, Italy, in April 1925 and by Mr. H. Twiddle in 1922, in which they recall conditions they witnessed in Peru in 1891. Excerpts from these letters appear in R. C. Murphy, "Oceanic and Climatic Phenomena along the West Coast of South America in 1925," *Geogr. Rev.* (1926): 26–54.

2. In the same way that a stone dropped into a pond excites waves that radiate away, so a pulse of winds over a limited part of the ocean excites waves that radiate away as undulations of the ocean surface or the thermocline. If the pulse persists for a few days at least, has spatial dimensions of a few hundred kilometers or more, and is located at the equator, then several types of waves that are possible only near the equator come into play.

The most prominent type is the Kelvin wave, which is bell-shaped in latitude, is centered on the equator, and travels swiftly and strictly eastward along that line. Upon reaching a coast, that of South America for example, the wave sends signals that travel poleward along the coast in both hemispheres. That is how Kelvin waves permit winds in the far western equatorial Pacific to affect oceanic conditions off California and Chile.

3. A group of experts who met in Princeton, N.J., in October 1982 to plan an international program to study El Niño was unaware that the most severe episode of the past century was occurring at the time. Satellites monitoring sea surface temperatures from space did report exceptionally warm waters in the eastern tropical Pacific, but the scientists chose to ignore those reports for reasons that seemed sound: an eruption of a volcano had injected into the atmosphere a huge amount of material, thus complicating the interpretation of satellite measurements and causing incorrect temperature readings. Instruments on unattended moorings were monitoring conditions in the tropical Pacific throughout 1982, but those measurements became available only after El Niño had come to an end, when a ship recovered the instruments.

4. K. Sponberg, "Weathering a Storm of Global Statistics," *Nature* (July 1, 1999): 13. K. Sponberg, *Compendium of Climatological Impacts* (Washington, D.C.: NOAA Office of Global Programs, 1999).

Chapter 3. A Construct of Ours

1. See note 1 of the prologue.
2. E. Hobsbawm, *The Age of Extremes* (New York: Pantheon Books, 1994).
3. Andrew Ross, *Strange Weather* (New York: Verso, New Left Books, 1991).
4. Chandra Mukerji, in *A Fragile Power: Scientists and the State* (Princeton, N.J.: Princeton Univ. Press, 1989).

Chapter 4. A Matchmaker

1. The TOGA (Tropical Oceans and Global Atmosphere) array of instruments has the following components: a network of buoys (the black dots on the map), moored to the ocean floor, with attached instruments that measure temperature and currents over the upper few hundred meters of the ocean; drifting buoys on the ocean surface (the arrows) that measure the temperature and the wind, and whose movements, tracked by satellites, yield information about surface currents; commercial ships that deploy instruments along their tracks (the extended lines) thus measuring temperature to a depth of a few hundred meters; tide gauges (the open circles) that measure sea level. The observing system also includes several geostationary and polar-orbiting satellites. Data from the array are available at the website *www.pmel.noaa.gov/toga-tao/*. See M. J. McPhaden et al. "The Tropical Ocean-Global Atmosphere Observing System: A decade of Progress," *J. Geophys. Res.* 103 (1998): 14169–240.

2. E. N. Lorenz, *The Essence of Chaos* (Seattle: Univ. of Washington Press, 1993).

3. G. I. Taylor, "Walker, Gilbert," Obituary Notice, *Q. J. R. Met. Soc.* 85: 186; Gilbert Thomas Walker (1868–1958), *Biographical Memoirs of the Fellows of the Royal Society* 8, (1962): 169–74.

4. J. N. Lockyer, "Simultaneous Solar and Terrestrial Changes," *Nature* 69 (1904): 351–57.

5. See note 3 above.

6. Sir G. Walker, "Correlations in Seasonal Variations of Weather," VIII, *India Meteor. Dept. Memoirs* 24, no. 4 (1923).

7. J. Tuzo Wilson, *IGY: The Year of the New Moons* (New York: Alfred A Knopf, 1961).

8. J. Holmboe, J. Namias, and M. G. Wurtele, "Jacob Bjerknes (1897–1975)," *Bull. Am. Met. Soc.* (1975): 1089–90.

9. E. Wenk, *The Politics of the Ocean* (Seattle: Univ. of Washington Press, 1972).

10. See note 9 above.

11. These words appear in a paper, circulated in 1954, by H. M. Stommel, "Why do our ideas about the ocean circulation have such a peculiarly dream-like quality? Or examples of types of observations that are badly needed to test oceanographic theories," in *Collected works of Henry M. Stommel* (Boston: Am. Meteorol. Soc., 1995).

12. D. F. Leipper, and J. M. Lewis, "Letter Exchange Documents 50 Years of Progress in Oceanography," *EOS* (Trans. Amer. Geophys. Union) no. 81, 18 (May 2, 2000).

Chapter 5. The Two Incompatible Cultures

The epigraph to this section is from Isaiah Berlin, "On Political Judgment," *The New York Review of Books* 43, no. 15 (October 3, 1996).

1. C. P. Snow, *The Two Cultures* (Cambridge: Cambridge Univ. Press, Canto Edition, 1993).

2. Gregory Jane and Steve Miller, *Science in Public* (New York: Plenum Trade, 1998).

3. For instance, on December 4, 1998, on the *Jim Lehrer News Hour* of the Public Broadcasting System, a scientist assured viewers that global warming is a matter of grave concern. The very next evening, on the same program, a businessman said the exact opposite and flatly rejected "the theory" of global warming.

4. R. K. Merton, *On the Shoulders of Giants* (New York: Harcourt, Brace, Jovanovich, 1985).

5. See note 1 above and also Isaiah Berlin, *The Crooked Timber of Humanity* (New York: Alfred A. Knopf, 1991); Isaiah Berlin, "My Intellectual Path," *New York Review of Books* 45, no. 8 (May 14, 1998).

6. See note 4 above.

7. See note 5 above.

8. Alan Lightman, "Looking Back at Pure World of Theoretical Physics," *New York Times*, May 9, 2000, p. F5.

9. G. di Lampedusa, *The Leopard* (New York: Pantheon Books, 1960).

10. Charles P. Kindleberger, *Manias, Panics, and Crashes* (New York: John Wiley & Sons, 1996); John Carswell, *The South Sea Bubble* (London: Alan Sutton Publishing, 1993).

11. The cost of a certain increase in carbon dioxide levels depends strongly on the length of time over which the increase occurs. A sudden change will be far more expensive than a gradual one that gives us time to adjust. The geological record has evidence of rapid climate changes in the past, changes that occurred within a decade or two. Allowing for such possibilities increases the estimated cost of future global warming significantly.

12. C. Azar, "Are Optimal CO_2 Emissions Really Optimal?" *Environmental and Resource Economics* 11, nos. 3–4: 301–15.

13. G. Bingham, R. Bishop, M. Brody, D. Bromley, E. Clark, W. Cooper, R. Constanza, T. Hale, G. Hayden, S. Keller, R. Norgaard, B. Norton, J. Payne, C. Russel, and G. Suter "Issues in Ecosystem Evaluation: Improving Information for Decision-Making," *Ecological Economics* 14 (1995): 73–90.

Chapter 6. "Small" Science versus "Big" Science

1. Derek J. de Solla Price, *Little Science, Big Science . . . and Beyond* (New York: Columbia Univ. Press, 1986).

2. John Suppe, "Exponential Growth of Geology, Mathematics, and the Physical Sciences for the Last Two Hundred Years and Prospects for the Future," lecture given at the 100th anniversary of Nanjing University, May 20–22, 2002. This paper documents a decrease, since the 1970s, in the rate of growth of the number of scientists and relates it to a similar decrease in the rate of growth of the number of people attending colleges and universities in the United States.

3. See note 1 above.

4. See note 1 above.

5. See note 2 above.

Chapter 7. The Perspective of a Painter

1. The discussion in this chapter draws heavily on E. H. Gombrich's superb book *Art and Illusion* (Princeton, N.J.: Princeton Univ. Press, 1960).

2. See note 1 above.

3. W. Metzger, Gesetze des sehens (Frankfurt: Verlag Waldemat Kramer, 1953.)

4. J. Gage, *Color and Culture* (Berkeley: Univ. of California Press, 1993); J. Gage, *Colour and Meaning: Art, Science, and Symbolism* (London: Thames and

Hudson, 1999); M. Pastoureau, *Blue: The History of a Color* (Princeton, N.J.: Princeton Univ. Press, 2001).

5. The confusion is in part attributable to the Renaissance scholar Georges Philander, who misidentified a primary color in *Decem Libros M. Vitruvius Pollionis de Architectura Annotationes* (Rome, 1544).

6. S. Zeki, *A Vision of the Brain* (Oxford: Blackwell Scientific Publications, 1993); S. Zeki, *Inner Vision: An Exploration of Art and the Brain* (Oxford: Oxford Univ. Press, 1999); S. Zeki, "Artistic Creativity and the Brain," *Science* 293: 51–51.

7. See chapter 17 for a further discussion of these topics.

8. H. Honour, *Romanticism* (New York: Harper and Row, 1979).

9. V. Scully, *Architecture* (New York: St. Martin's Press, 1991).

10. Whether computer simulations contribute to the scientific understanding of natural phenomena has been a topic of heated debates. For a sober discussion see S. D. Norton and F. Suppe, "Epistemology of Atmospheric Modeling," in *Changing the Atmosphere: Expert Knowledge and Environmental Governance,* Paul N. Edwards and Clark A. Miller, eds. (Cambridge, Mass.: MIT Press, 2001).

Chapter 8. The Perspective of a Poet

1. The quotation in the epigraph to this chapter comes from Alfred Appell, "An Interview with Nabokov," *Wisconsin Studies in Contemporary Literature* 8 (spring 1967): 140–41.

2. The exceptions are thermal vents on the ocean floor, where exotic lifeforms thrive under very high pressures and temperatures.

3. For a sketch showing the various surface currents see figure 14.1. The related discussion in chapter 14 emphasizes that the cold water that rises to the surface in upwelling zones is from relatively shallow depths, from within the thermocline. Exchanges between the upper and very deep ocean are confined mainly to high latitudes.

4. See chapter 13 for a discussion of winds on a water-covered globe.

5. The Indian Ocean, because it is bounded by continents at low latitudes to the north, has strong monsoon winds.

6. In the eastern equatorial Atlantic, the presence of the West African land mass just north of the equator contributes significantly to the climate asymmetries. An interesting contrast is the Indian Ocean, which is bounded by land across its entire width in low northern latitudes but which does not have sea surface temperature patterns with pronounced asymmetries. In that ocean the winds are such that cold water does not appear at the ocean surface except in very small regions, so that ocean-atmosphere interactions are of secondary importance. See S. G. Philander et al., "Why the ITCZ Is Mostly North of the Equator," *J. Climate* 9 (1996): 2958.

Chapter 9. The Perspective of a Musician

1. G. Buzyna, R. C. Pfeffer, and R. Kung, *J. Fluid Mechanics* 145 (1984): 377–403.

2. R. Legeckis, "Long Waves in the Eastern Equatorial Pacific," *Science* 197 (1977): 1179–81.

3. See chapter 15 for more details.

Chapter 10. A Marriage of the "Hard" and "Soft" Sciences

1. E. N. Lorenz, *The Essence of Chaos* (Seattle: Univ. of Washington Press, 1993).

Chapter 11. The Cloud

1. The story of Luke Howard, who, in London in December 1802, introduced the currently used nomenclature for clouds, is told by R. Hamblyn in *The Invention of Clouds* (New York: Farrar, Straus, and Giroux, 2001). The English language has a very rich vocabulary but, before Howard, did not include many words to describe the great variety of clouds we observe in the sky. Howard therefore received immediate acclaim, and he stimulated several poets and painters to take a keen interest in clouds. Even the renowned Goethe wrote a poem "In Honour of Howard."

2. Richard Holmes, *Coleridge: Early Visions* (London: Hodder and Stoughton, 1989).

3. D. King-Hele, *Shelley: His Thought and Work* (London: MacMillan, 1970), 219–27; F. H. Ludlam, "The Meteorology of the Ode to the West Wind," *Weather* 27 (1972): 503–14.

Chapter 12. Predicting the Weather

The epigraph to this chapter comes from Sir William Thomson (Lord Kelvin), "The Practical Applications of Electricity," in *Popular Lectures and Addresses* (London: Macmillan, 1889) vol. 1, *Constitution of Matter*, 72–73.

1. R. A. Bryson, "Typhoon Forecasting, 1944, or, the Making of a Cynic," *Bull. Am. Met. Soc.* 81, no. 10 (2000): 2393–97. See also E. Durschmied, *The Weather Factor* (New York: Arcade Publishing, 2000).

2. R. Hamblyn, *The Invention of Clouds* (New York: Farrar, Straus, and Giroux, 2001).

3. Mark Monmonier, *Air Apparent* (Chicago: Univ. of Chicago Press, 1999).

4. W. B. Meyer, *Americans and Their Weather* (Oxford: Oxford Univ. Press, 2000).

5. Gisela Kutzbach, *The Thermal Theory of Cyclones* (Boston: American Meteorological Society, 1979).

6. N. A. Phillips, "Carl-Gustaf Rossby: His Times, Personality, and Actions," *Bull. Am. Met. Soc.* 79, no. 6 (1998): 1097–112.

7. Lewis F. Richardson, *Weather Prediction by Numerical Process* (Cambridge: Cambridge Univ. Press, 1922).

8. J. Smagorinsky, *The Beginnings of Numerical Weather Prediction and General Circulation Modeling: Early Recollections,* Advances in Geophysics, vol. 25 (New York: Academic Press, 1983).

9. Paul N. Edwards and Clark A. Miller, eds., *Changing the Atmosphere: Expert Knowledge and Environmental Governance* Cambridge, Mass.: MIT Press, 2001.

10. Phillip D. Thompson, "A History of Numerical Weather Prediction in the United States," *Bull. Am. Met. Soc.* 64 (1983): 755–69.

Chapter 13. Investigating the Atmospheric Circulation

1. E. N. Lorenz, *The Nature and Theory of the General Circulation of the Atmosphere* (Geneva: World Meteorological Organization, 1967); J. P. Peixoto and A. H. Oort, *Physics of Climate* (New York: American Institute of Physics, 1992).

2. For an assessment of the climate models see Houghton et al. eds., *Climate Change 2002 (The Scientific Basis)* (Cambridge: Cambridge Univ. Press, 2002).

3. N. A. Phillips, "The General Circulation of the Atmosphere: A Numerical Experiment," *Q. J. R. Meteorol. Soc.* 82 (1956): 123–64.

4. See note 2 above.

5. J. S. Fein and P. L. Stephens, eds., *Monsoons* (New York: John Wiley & Sons, 1987).

6. G. Simpson, "The Southwest Monsoon," *Q. J. R. Met. Soc.* 199, no. 17 (1921): 150–73.

7. Quoted in R. Proctor, "Sunspot, Storm, and Famine," *Gentleman's Magazine*, December 1877, 701.

8. Sir G. Walker, "Correlations in Seasonal Variations of Weather," VIII, *India Meteor. Dept. Memoirs* 24, no. 4 (1923).

9. J. Bjerknes, "A Possible Response of the Atmospheric Hadley Circulation to Equatorial Anomalies of Ocean Temperature," *Tellus* 18 (1966): 820–29.

10. The data available in the late 1970s were inadequate for a detailed description of a specific El Niño. E. M. Rasmussen and T. H. Carpenter therefore described a "canonical" or "composite" El Niño that starts with the appearance of unusually warm water off the coast of Peru in March and April and then develops in a westward direction. See *Mon. Wea. Rev.* 110 (1982): 354–84. A few events in the 1960s and 1970s did develop in such a manner, but El Niño of 1982 started in the west and developed in an eastward direction. From the data available today, it is clear that each El Niño is distinct. What they all have in common is a warming of the eastern equatorial Pacific, where the thermocline deepens, and a relaxation of the trade winds along the equator.

11. J. Shukla and J. G. Charney, "Predictability of the Monsoons," in *Monsoon Dynamics,* J. Lighthill and R. Pierce, eds. (Cambridge: Cambridge Univ. Press, 1981); N. C. Lau, "*Modeling the Seasonal Dependence of the Atmospheric Response to Observed El Niño Episodes, 1962–1976,*" *Mon. Wea. Rev.* 113 (1985): 1970–96.

12. J. M. Wallace and D. S. Gutzler, "Teleconnections in the Geopotential Height Field during the Northern Hemisphere Winter," *Mon. Wea. Rev.* 109 (1981): 784–812.

13. K. K. Kumar, B. Rajagopalan, and M. A. Cane, "On the Weakening Relationship between the Indian Monsoon and ENSO," *Science* 284, no. 5423 (1999): 2156–59.

Chapter 14. Exploring the Oceans

1. Henry Ellis, "A Letter to the Rev. Dr. Hales, F.R.S., from Captain Henry Ellis, F.R.S., Dated Jan. 7, 1750–51, at Cape Monte Africa, Ship *Earl of Halifax,*" *Philosophical Transactions (1683–1775)* 47 (1751–52): 211–16.

2. R. Kunzig, *The Restless Sea* (New York: W. W. Norton & Co., 1999).

3. Susan Schlee, *The Edge of an Unfamiliar World: A History of Oceanography* (New York: E. P. Dutton & Co., 1973), and see note 2 above.

4. H. Stommel, "The Westward Intensification of Wind-Driven Ocean Currents," *Trans. Am. Geophys. Union* 29 (1948): 202–6.

5. R. Seager, D. S. Battisti, J. Yin, N. Naik, A. C. Clement, and M. A. Cane, "Is the Gulf Stream Responsible for Europe's Mild Winters?" *Q. J. R. Meteorol. Soc.* 127 (2001).

6. E. Wenk, *The Politics of the Ocean* (Seattle: Univ. of Washington Press, 1972).

7. W. Duing, P. Hisard, E. Katz, J. Meincke, L. Miller, K. Moroshkin, G. Philander, A. Rybnikov, K. Voigt, and R. Weisberg, "Meanders and Long Waves in the Equatorial Atlantic," *Nature* 257 (1975): 280–84.

8. R. Legeckis, "Long Waves in the Eastern Equatorial Pacific," *Science* 197 (1977): 1179–81.

9. S. G. Philander, "Instabilities of Zonal Equatorial Currents," *J. Geophys. Res.* 81, no. 21 (1976): 3725–35.

10. C. Wunsch and A. E. Gill, "Observations of Equatorially Trapped Waves in Pacific Sea Level Variations," *Deep Sea Res.* 23 (1976): 371–90.

11. T. Matsuno, "Quasi-geostrophic Motion in Equatorial Areas," *J. Meteorol. Soc. Japan* 2 (1966): 25–43; D. W. Moore, Ph.D. diss., Harvard University, 1969.

12. G. Veronis and H. Stommel, "The Action of Variable Wind-Stresses on a Stratified Ocean," *J. Mar. Res.* 15 (1956): 43–69.

13. Seafarers have known since the Middle Ages at least that the Somali Current reverses direction seasonally. See B. Warren, "Medieval Arab References to

the Seasonally Reversing Currents of the North Indian Ocean," *Deep Sea Res.* 13 (1965): 167–71. The first attempt to explain the generation of that current after the onset of the monsoons was by M. J. Lighthill, "Dynamic Response of the Indian Ocean to the Onset of the Southwest Monsoon," *Philos. Trans. R. Soc. London*, ser. A, (1969): 265 45–93.

14. K. Wyrtki, "Teleconnections in the Equatorial Pacific Ocean," *Science* 180 (1973): 66–68.

15. D. Halpern, "Observations of Annual and El Niño Thermal and Flow Variations along the Equator at 110W and 95W during 1980–1985," *J. Geophys. Res.* 92 (1987): 8197–212.

16. K. Yoshida, "A Theory of the Cromwell Current and Equatorial Upwelling," *J. Oceanogr. Soc. Japan* 38 (1959): 215–224; J. McCreary, "Eastern Tropical Ocean Response to Changing Wind Systems with Application to El Niño," *J. Phys. Oceanogr.* 6 (1976): 632–45; J. O'Brien and H. Hurlburt, "Equatorial Jets in the Indian Ocean," *Science* 124 (1974): 1075–77.

17. M. A. Cane and E. S. Sarachik, "Forced Baroclinic Ocean Motion I," *J. Marine Res.* 34, no. 4 (1976): 629–65.

18. K. Bryan, "A Numerical Method for the Study of the World Ocean," *J. Comput. Phys.* 4 (1969): 347–76.

19. S. G. Philander and A. Seigel, "Simulation of El Niño of 1982–83," in *Coupled Ocean-Atmosphere models*, J. Nihoul, ed. (Amsterdam: Elsevier, 1985), 517–41.

20. S. Hayes, L. Mangum, J. Picaut, A. Sumi and K. Takeuchi, "TOGA-TAO: A Moored Array for Real-Time Measurements in the Tropical Pacific Ocean," *Bull. Am. Met. Soc.* 72 (1991): 339–47.

21. A. Leetmaa and M. Ji, "Operational Hindcasting of the Tropical Pacific," *Dyn. Atmos. and Oceans* 13 (1989): 465–90.

Chapter 15. Reconciling Divergent Perspectives on El Niño

1. A collection of articles that review recent scientific research related to El Niño, each with a comprehensive list of references, appears in vol. 103, no. C7, of the *Journal of Geophysical Research* (June 29, 1998). For a summary of scientific research up to 1990 see S. G. Philander, *El Niño, La Niña, and the Southern Oscillation"* (New York: Academic Press, 1990). See also S. G. Philander and A. V. Fedorov, "Is El Niño Sporadic or Cyclic?" *Annual Reviews of Earth and Planetary Sciences* 31 (2003): 579–94.

2. J. Bjerknes, "A Possible Response of the Atmospheric Hadley Circulation to Equatorial Anomalies of Ocean Temperature," *Tellus* 18 (1966): 820–29; J. Bjerknes, "Atmospheric Teleconnections from the Equatorial Pacific," *Mon. Wea. Rev.* 97 (1969): 163–72.

3. K. Wyrtki, "El Niño — The Dynamic Response of the Equatorial Pacific to Atmospheric Forcing," *J. Phys. Oceanogr.* 5 (1975): 572–84.

4. K. Wyrtki, E. Stroup, W. Patzert, R. Williams, and W. Quinn, "Predicting and Observing El Niño," *Science* 191 (1976): 343–46.

5. S. G. Philander, T. Yamagata, and R. C. Pacanowski, "Unstable Air-Sea Interactions in the Tropics," *J. Atm. Sci.* 41 (1984): 604–13; T. Yamagata, "Stability of a Simple Air-Sea Coupled Model in the Tropics," in *Coupled Ocean-Atmosphere Models,* J. C. J. Nihoul, ed., Elsevier Oceanogr. Ser. 40 (New York: Elsevier, 1985), 637–57.

6. M. J. McPhaden and X. Yu, "Equatorial Waves and the 1997–98 El Niño," *Geophys. Res. Lett.* 26 (1999): 2961–64.

7. S. E. Zebiak and M. A. Cane, "A Model El Niño–Southern Oscillation," *Mon. Wea. Rev.* 115 (1987): 2262–78.

8. P. S. Schopf and M. J. Suarez, "Vacillations in a Coupled Ocean-Atmosphere Model," *J. Atm. Sci.* 45 (1988): 680–702; D. S. Battisti and A. C. Hirst, "Interannual Variability in the Tropical Ocean-Atmosphere System," *J. Atm. Sci.* 46 (1989): 1687–712.

9. J. D. Neelin, "A Hybrid Coupled General Circulation Model for El Niño Studies," *J. Atm. Sci.* 47 (1990): 674–93.

10. P. Chang, L. Ji, H. Li, and M. Flugel, "Chaotic Dynamics versus Stochastic Processes in El Niño–Southern Oscillation in Coupled Ocean-Atmosphere Models, *Physica D* 98 (1996): 301–20; C. J. Thompson and D. S. Battisti, "A Linear Stochastic Dynamical Model of ENSO. Part I: Model Development," *J. Climate* 13 (2000): 2818–32, and "Part II: Analysis," *J. Climate* 14 (2001): 445–66; C. Eckert and M. Latif, "Predictability of a Stochastically Forced Hybrid Coupled Model of El Niño," *J. Climate* 10 (1997): 1488–1504; B. Blanke, J. D. Neelin, and D. Gutzler, "Estimating the Effect of Stochastic Wind Stress Forcing on ENSO Irregularity," *J. Climate* 10 (1997): 1473–86, M. S. Roulston and J. D. Neelin, "The Response of an ENSO Model to Climate Noise, Weather Noise, and Intraseasonal Forcing," *Geophys. Res. Lett.* 27 (2000): 3723–26.

11. See Philander and Fedorov in note 1 above.

12. J. D. Neelin, "The Slow Sea Surface Temperature Mode and the Fast Wave Limit," *J. Atm. Sci.* 48 (1991): 584–606. For a discussion of the background conditions that favor some modes rather than others see Philander and Fedorov in note 1 above.

13. See note 10 of chapter 1 for a discussion.

14. M. Latif, K. Sperber, J. Arblaster, and P. Braconnot, "ENSIP: El Niño Simulation Intercomparison Project, *Climate Dynamics* 18 (2001): 255–76.

Chapter 16. Taking a Long-Term Geological View

1. For an excellent review of climate changes during the Cenozoic see J. Zachos, M. Pagani, L. Sloan, E. Thomas, and K. Billups, "Trends, Rhythms and Aberrations in Global Climate 65 Ma to the Present," *Science* 292 (2001): 683–93.

2. J. R. Petit et al., "Climate and Atmospheric History of the Past 420,000 Years from the Vostok Core, Antarctica," *Nature* 399 (1999): 429–36. By drilling into a glacier, scientists can recover ice that fell as snow thousands of years ago. Trapped in the ice are air bubbles that tell us about the atmospheric composition in the past. The temperature variations are inferred from variations in the concentration of certain hydrogen isotopes in the ice. (The highly correlated curves are therefore from entirely separate measurements.)

3. P. B. deMenocal, *Plio-Pleistocene African Climate, Science* 270, (1995): 53–59.

4. J. Imbrie and K. P. Imbrie, *The Ice Ages: Solving the Mystery* (Cambridge, Mass.: Harvard Univ. Press, 1979).

5. J. Kennett, G. Keller, and M. Srinivasan, *Mem. Geol. Soc. Amer.* 163 (1985): 197–236, analyze cores from the equatorial Pacific and show that 22, 16, and even 8 million years ago, in the eastern equatorial Pacific, the thermocline was below the photic zone, significantly deeper than it is today. A. C. Ravelo, D. H. Andreasen, M. W. Wara, M. Lyle, and A. Olivarez, in "California Margin (Tanner Basin) Records of Plio-Pleistocene Circulation and Climate" (Abstract from fall meeting of the Am. Geophys. Union, 2001), find that, up to 3 million years ago, surface waters were warm off California but temperatures then started to fall gradually to their present low values. J. R. Marlow, C. B. Lange, G. Wefer, and A. Rosell-Mele, in "Upwelling Intensification as Part of the Pliocene-Pleistocene Climate Transition," *Science* 290 (2001): 2288–94, show that cold surface water first appeared in the intense upwelling zone off Southwest Africa around 3 million years ago. P. Molnar and M. Cane, in *Paleoceanogr.* 17 (2002), describe records that indicate permanent El Niño conditions earlier than 3 million years ago.

6. S. G. Philander and A. V. Fedorov, "The Role of the Tropics in Changing the Response to Milankovich Forcing Some Three Million Years Ago," *Paleoceanography* 18, no. 2 (2003).

Chapter 17. Famines in India

1. Mike Davis, *Late Victorian Holocausts: El Niño, Famines, and the Making of the Third World.* (New York: Verso, 2000).

2. Parliamentary papers 1881, 68, "Famine Commission — Financial statement." Quoted in Sheldon Watts, *Epidemics and History: Disease, Power, and Imperialism* (New Haven, Conn.: Yale Univ. Press, 1997).

3. Adam Smith, *An Inquiry into the Nature and Causes of the Wealth of Nations* (1776).

4. Norman J. Lockyer and W. Hunter, *Sun Spots and famines* The Nineteenth Century, Nov. 1877.

5. Sen Amartya, *Poverty and Famines* (Oxford: Clarendon Press, 1981); J. Dreze and Sen Amartya, *Hunger and Public Action* (Oxford: Clarendon Press, 1989).

6. A. Sen, "Apocalypse Then," *New York Times Book Review*, February 18, 2001.

7. M. Massing, "Does Democracy Avert Famine?" *New York Times*, March 1, 2003, p. B7.

Chapter 18. Fisheries of Peru

1. R. Jordan, "Biology of the anchoveta," in *Proceedings of the Workshop on the Phenomenon Known as El Niño* (Paris: UNESCO, 1980), 249–76.

2. J. B. C. Jackson et al., "Historical Overfishing and the Recent Collapse of Coastal Ecosystems," *Science* 293 (2001): 629–38.

3. M.-E. Carr and K. Broad, "Satellites, Society, and the Peruvian Fisheries during the 1997–1998 El Niño," in *Satellites, Oceanography, and Society*, D. Halpern, ed. (New York: Elsevier, 2000); L. A. Hammergren, "Peruvian Political and Administrative Responses to El Niño," in *Resources Management and Environmental Uncertainty*, vol 11, M. H. Glantz and J. D. Thompson, eds. (New York: John Wiley and Sons, 1981).

Chapter 19. Droughts in Zimbabwe

1. The characters in this chapter are all fictional.

Epilogue: Becoming Custodians of Planet Earth

1. "The Gambler," by Don Schlitz, Copyright © 1977 Sony/ATV Tunes LLC d/b/a Cross keys Publishing Co.

2. P. D. Ward and D. Brownlee, *Rare Earth* (New York: Copernicus, 2000).

3. E. D. Nering "*The Mirage of a Growing Fuel Supply*," *New York Times*, June 5, 2001.

4. The *New York Times* of June 8, 2001, reported that Du Pont aims to cut its emissions by 40 percent, that Alcoa plans reductions of 25 percent, and that the Mexican oil giant Pemex also plans carbon dioxide reductions.

INDEX

Accademia del Cimento, 165

Agulhas Current, 200

Alain, 94, 95, 115, 117

angel, 12, 28, 33

años de abundancia, 28

Antarctic Circumpolar Current, 200, 202

Arrhenius, 35, 179

Atlantis, 41

Bacon, Francis, 227

Benguela Current, 197, 200

Bergen school, 168, 169

Berlin, Isaiah, 65, 70, 263

Bjerknes, Jacob, 49, 51, 54, 56, 85, 104, 120, 168, 186, 214–16, 263, 267, 269

Boltzmann, Ludwig, 101

Brandes, Heinrich Wilhelm, 165

Braque, Georges, 99

Calypso, 41

Captain Ahab, 19

catastrophists,16

Cenozoic, 228, 231, 270

Challenger, 41, 42, 43, 69, 77, 80, 301

Charles, Jacques, 164

Charney, Jule, 171, 172, 267

Cold War, 37, 60

Coleridge, Samuel, 152

Confucius, 17

Constable, John, 93, 95, 97, 100, 105

Corot, Jean-Baptiste-Camille, 97

Cretaceous, 59, 89

Currents: Agulhus, 200; Antarctic Circumpolar Current, 200, 202; Benguela, 197, 200; Equatorial Undercurrent, 52, 53, 209, 210; Gulf Stream, 42, 52, 124, 134, 135, 149, 192–200, 206, 233; Kuroshio, 42, 124, 134, 196–98, 233; Malvinas, 200

da Vinci, Leonardo, 96, 100

Darwin, Charles, 131, 142, 194

de Solla Price, 82

Delacroix, Eugène, 98, 110, 140

Democritus, 106

Derain, André, 100

di Lampedusa, 72, 264

Dickens, Charles, 13

doldrums, 11, 182

Dove, Heinrich, 165

Duchamps, Marcel, 112

Dust Bowl, 36

Ecclesiastes, 15, 16, 240, 244

Einstein, Albert, 68, 85, 100, 255

Ekman, Vagn Walfrid, 196

El Viejo, 30

ENSO, 12, 21, 260, 268

Equatorial Undercurrent, 52, 53, 209, 210

Espy, James Pollard, 166, 167, 168

Euclid, 67

European Center for Medium Range Weather Forecasting, 175

Exponential growth, 255, 256

Fourier, Jean-Baptiste, 179

Franklin, Benjamin, 193

Galápagos Islands, 207, 223

Geophysical Fluid Dynamics Laboratory, 173, 210

Global Atmospheric Research Program, 52

Global Positioning System, 56
Gradgrind, 13
guano, 28
Gulf Stream, 42, 52, 124, 134, 135, 149, 192–200, 206, 233

Hadley, George, 180, 181, 184, 188, 267, 269
Hamlet, 11, 113, 114
Hard Times, 13
Hardy, Thomas, 44
Haydon, Benjamin, 140
Henry IV, Part I, 213
Herrick, Robert, 139, 150
Hugo, Victor, 237
Hutton, James, 16, 260
Huxley, Thomas, 142

Ice Ages, 59, 89, 137, 230, 231, 254, 271
IGY. *See* International Geophysical Year
Ingres, Jean-Auguste-Dominique, 110
Integrated Assessment Models, 77
International Decade of Ocean Exploration, 52, 204
International Geophysical Year, 49, 52
invisible college, 86

Keats, John, 140
Keeling, Charles, 179
Kelvin, 14, 57, 208, 215, 262, 266
Keynes, Maynard, 256
Kuroshio, 42, 124, 134, 196–98, 233
Kyoto Protocol, 77

Leopard, 72, 264
Lightman, Alan, 72, 264
Lockyer, Norman, 47, 185, 263
Longfellow, Henry Wadsworth, 52
Lorenz, Edward, 46, 143, 144, 146, 263, 266, 267
Lorrain, Claude, 97
Louis XIV, 110

Macbeth, 78
Malvinas Current, 200

Mars, 106, 107, 227
Massachusetts Institute of Technology, 46, 54, 217
Matisse, Henri-Emile-Benoît, 100, 114
Maury, Matthew, 193,
Mayor of Casterbridge, The, 44
Melville, Herman, 42, 189
Menuhin, Yehudi, 129
Mercury, 69
Meteor, 41
Meteorological Society of the Palatinate, 165
Midsummer Night's Dream, A, 120
Milankovich, Milutin, 230, 271
Moby-Dick, 19, 42, 189
moguls, 143, 144
Montgolfier brothers, 164

Nabakov, Vladimir, 118, 125, 265
Neptune, 69
New York Times, 68
Newton, Isaac, 68, 69, 73, 74, 84, 98, 100, 102, 103, 140, 143, 230

Ode to the West Wind, 133, 146, 266

Pareto's Law, 85
Pascal, Blaise, 164
Phillips, Norman, 183, 267
phytoplankton, 123, 240
Picasso, Pablo, 99
Pizarro, Francisco, 121
Pope, Alexander, 69
Pygmalion, 114

Reader's Digest, 31
Redfield, William, 166, 167
Rich, Barnaby, 81, 64
Richard III, 27
Richardson, L. F., 170–72, 267
Rogers, Kenny, 251
Rossby, Carl Gustav, 169, 208, 267
Royal Society of London, 82, 86
Ruskin, John, 118, 119

Shakespeare, William, 11, 27, 78, 113, 114, 120, 140, 213
Shelley, Percy Bysshe, 133, 146, 147, 151, 152, 266
Smagorinsky, Joseph, 173, 267
Smithsonian Institution, 167
Snow, C.P., 65, 263
Soybean Digest, 31
Spice Girls, 38
Sputnik, 51, 52, 173, 204
Stommel, Henry, 53, 198, 199, 206, 208, 210, 263, 268

telegraph, 45, 54, 167
Thompson, Benjamin, 190, 191, 266
Time, 36
TOGA, 43, 56, 60, 61, 87, 204, 211, 262
Twelfth Night, 28

uniformitarians, 16
Ussher, James, 16

Venus, 106, 107, 157, 227
Vermeer, Jan, 112
Velázquez, Diego, 112
von Guericke, Otto, 164
von Humboldt, Alexander, 58, 59
von Neumann, John, 171, 173
Vulcan, 69

Walker, Gilbert, 47–49, 60, 85, 118, 186, 187, 237, 260, 263, 267
Wall Street, 11, 38
Waves: Kelvin, 14, 57, 208, 215, 262; Rossby, 208
World Meteorological Organization, 52, 172, 173, 267
Wyrtki, Klaus, 207, 214, 269, 270

dehmitt